AUSTRALIAN TRACTO

INDIGENOUS TRACTORS AND SE
MACHINES IN RURAL AU

Dr Graeme Quick, born and educated in Australia, has broad international experience in engineering for agriculture — in teaching, research, extension and consulting. He is the author of over 140 publications, including the key reference book (with W.F. Buchele), *The Grain Harvesters*, published by the American Society of Agricultural Engineers. He holds 14 patents on farm equipment developments, from which several designs are in production and use worldwide. He has designed many different types of farm equipment and has personally built several small tractors and self-propelled machines. A large amount of his professional career has been in engineering in 35 other countries.

In 1983 Graeme Quick was awarded the Urrbrae Medal for an outstanding contribution to Australian Agriculture. In 1988 he was made a Fellow of the Institution of Engineers, Australia, and in 1995 a Fellow of the American Society of Agricultural Engineers. In 2004 ASAE awarded him the C.H. McCormick–J.I. Case Gold Medal for his professional contributions.

Married with three sons, he and his wife Marlene live in Queensland, where he is an engineering consultant and continues technical writing.

A Jelbart Crude Oil Tractor was one of those tested in the 1918 Werribee tractor trials — Australia's original tractor tests. As part of the tests the tractors were belt tested on the University of Melbourne's engine dynamometer. Here the Jelbart is seen parked behind the University's School of Engineering in 1918. (Photo courtesy Victorian Department of Agriculture)

Turning a ticklish situation to advantage, this farmer loads his Fordson, complete with Howard Rotavator, on a Chevrolet truck. (J.P. Richards photo, c. 1930s)

AUSTRALIAN TRACTORS

INDIGENOUS TRACTORS AND SELF-PROPELLED MACHINES IN RURAL AUSTRALIA

Graeme R. Quick

ROSENBERG

This edition first published in Australia in 2006
by Rosenberg Publishing Pty Ltd
PO Box 6125, Dural Delivery Centre NSW 2158
Phone: 612 9654 1502 Fax: 612 9654 1338
Email: rosenbergpub@smartchat.net.au
Web: www.rosenbergpub.com.au

Front cover pictures
Top right: 1917 Jelbart owned by Jack Burrell, Gilgandra, NSW.
Top left: 2005 Dingo K9-3 mini-loader made in Dalby, Queensland (courtesy Dingo Mini-Diggers Pty Ltd).
Centre: 1917 Sunshine Model A of Dave Berthelson, Gayndah, Queensland.
Bottom: Jeff Morgan's tandem Chamberlain rig (courtesy *The Old Machinery Magazine*).

Set in 11 on 13 point Adobe Garamond Pro
Printed in China by Everbest Printing Co Limited
Reprinted in 2007

The National Library of Australia Cataloguing-in-Publication data

Quick, Graeme R.
Australian tractors : indigenous tractors and self-propelled machines.

New ed.
Bibliography.
Includes index.
ISBN 9781877058394.

1. Tractors - Australia - History. 2. Tractor industry - Australia - History. I. Title.

629.22520994

CONTENTS

FOREWORD

As curator of one of Australia's major collections of agricultural machinery, I've been encouraged over the past decade by the strong growth of interest in old tractors.

Once-derelict machines have been rescued from graveyards in back paddocks and under peppercorn trees and have been preserved or painstakingly restored to operating condition. Today many can be seen at museums and rallies around Australia, reminding us of the significant role they played in Australia's rise to nationhood.

Recently, perceptive enthusiasts and collectors (both public and private) have been focusing their attention on tractors manufactured in Australia but have been handicapped by a debilitating lack of information on the subject.

Graeme Quick's book brings a welcome change to this situation. Here is a reference that at last brings together details on all the tractors known to have been designed or built by Australians. The story of each machine has been painstakingly researched and includes much technical data for the mechanically minded.

The story of Australian tractors is more than just technological, however. It is intertwined with changing economic and social conditions, and with the lives of a wide range of Australians — farmers, engineers, inventors, manufacturers, academics, business people, and others. It also demands sober analysis of the unique contributions, achievements and failures of Australians, viewed from the wider context of world developments in tractor design and manufacture.

It is a mark of the author's knowledge, research efforts and scholarship that he has been able to bring all these threads together to weave the story into a rich tapestry that will interest and inform a wide range of readers, not only tractor enthusiasts.

It has been a special pleasure for me to work with Graeme on this important topic and contribute to his book. I hope each reader will derive as much enjoyment from its pages as I did from my involvement.

***

Fifteen years after writing the above I'm delighted that persistent interest in the history of Australian tractors has resulted in a new edition of Graeme's book. I'm also pleased to see that the original contents have withstood scrutiny over the years — only minor changes and many useful additions are to be found in the historical parts to create this new edition. The concluding segments bring fresh data, many never before seen pictures and valuable new reference materials that are both useful and enlightening.

As we move into the 21st century, finding ways to produce food within sustainable economies is a major challenge for our global society. It is fascinating to see that many of the stories and issues raised in this book — about creativity, innovation, enterprise and local manufacture — are even more important for Australians now than when the book first appeared.

Martin L. Hallett, 2005
Previously Senior Curator, Science and Technology,
Museum of Victoria

ACKNOWLEDGMENTS

The preparation of this book was greatly assisted by the cooperation of the following individuals and organisations, whose help is gratefully acknowledged: Bill Brown, Agricultural Engineering Centre, Werribee; Brian Burke, Killara; A.H. (Bob) Chamberlain, Toorak; Peter Cousins, Henry Ford Museum, Dearborn, Michigan; John Creasey, Whiteknights, Reading, England; Russ Devlin, International Harvester, Dandenong; Brian Dunn and Vin Delahunty, Tractor and Machinery Association, Melbourne; Don Quick, Lara; Frank Gason, Ararat; Ray Gillett, Ringwood; Doug Goddard, Tamworth; Peter Griffiths and Barry Waghorn, Chamberlain John Deere; Alex and Peter Hayward, Camden; Klaus Herrmann, Hohenheim, Germany; Ray Hooley, Lincoln, England; Noel Howard, Ford Australia, Campbellfield; Lloyd Hughes, formerly with *Power Farming*, Melbourne; Bruce McDonald, Canberra; Neal McDonald, Frankston; John Parncutt, Melbourne; Pen Richards, Killara; Les Schulz, Horsham; Margaret and Phil Simpson, Roseville; Ian Stewart, The Olde Machinery Mart, Rockhampton; Lance Wilson, *The Old Machinery Magazine*, Port Macquarie; Harry Sweeting, St Ives; Gil Tippett, Horsham; Carl Upton, Toowoomba; James Nagorcka, Tarrington, Victoria; Philip Wyndham, Tracmach, Rossmoyne, WA; Jeff Morgan, Karoonda, SA; Gary Briggs, Dalby; Theo Reinhold, Brisbane; Terry Buckaloo, Sterling , Illinois.

There were many others who have helped behind the scenes in various ways, especially members of the staff of the State Library of New South Wales, the Australian Institute of Petroleum, the Science Museum, London; the New South Wales Department of Agriculture, Campbelltown Steam Museum members and, above all, Martin Hallett in Melbourne, his wife Liza Dale-Hallett and colleagues of the Museum of Victoria. Finally, special thanks to my wife, Marlene, who for years has had to tolerate books and papers at times in the wrong places at home. She is an excellent editor and has always had good advice in the choice of words for writing on tractors and farm equipment.

The author invites comments from anyone willing to draw attention to any omissions or mistakes that may have occurred in this latest edition. There is always more to learn. Every effort will be made to include any relevant new knowledge in the next edition.

Graeme R. Quick
grquick@hypermax.net.au

1 TRACTORS: AN AUSTRALIAN SETTING

Man is worth what man produces, and when machines increase his production, he is worth more.
— *R.G. Le Tourneau, in* Mover of Men and Mountains, *1972*

Australia has some of the world's largest and most labour-efficient farms. A well-equipped wheat farm can produce a tonne of wheat from start to finish with a labour input of less than one man-hour. Tractors are an essential part of that industry. The average Australian farm has three or more tractors.

Isolation, environmental extremes, shallow soils, minimal government assistance, and widely fluctuating economic circumstances over the years have forced farmers Down Under to operate at high levels of efficiency.

Since the earliest colonial days, the wellbeing of the rural economy has been dependent on its ability to export. Agricultural commodities started out as the backbone of the nation's economy. There was even a time when the country was said to ride on the sheep's back.

Throughout the industrialised world, mechanisation has become a cornerstone of production agriculture and the tractor is the modern symbol of farm mechanisation. Australian farmers have always been prepared to adapt technologies to their needs and to participate in the development of suitable machines in order to compete with the rest of the world.

This book is the story of Australian tractor makers and their machines. It is intended to put into perspective such questions as:

- How did the indigenous tractor industry get started?
- How extensive was the Australian tractor industry?
- Where did horse or steam power fit into Australian broadacre farming?
- Was the industry of any national consequence?
- Did Australian tractor makers contribute to worldwide tractor developments?
- Where has the tractor industry and market been heading?

Tractor engineering developments are outlined with reference to an Australian setting.

The story of the 45 or so makers of tractors and self-propelled agricultural machines detailed here is meant to stimulate or energise collectors — and evoke fond memories among those who have put in some time down on the farm.

Origin of the Name 'Tractor'

The word 'tractor' was first coined commercially in 1906 by Messrs Hart and Parr, pioneer engine builders of Charles City, Iowa, or rather, by their sales manager. He was puzzling over an advertisement he was writing. The title 'gasoline traction engine' then in use to describe the Hart-Parr machine (and others) was too unwieldy. He wrote the word 'tractor' into the ad and that name stayed with the machine. Hart-Parr tractors were among the first internal combustion (I-C) engined farm tractors to be designed specifically for drawbar work.

Perhaps unbeknown to sales manager Williams however, George H. Edwards in Chicago had patented a 'tractor' in 1890. Edwards deserves mention for being first to patent the word for agriculture. However, Edwards' design was a steamer and there is no evidence that his design was ever commercialised.

Charles Hart and Charles Parr were engineering students in Madison, Wisconsin, when they made their first two-cylinder engine in 1898 as an honours project. They found backing to set up a factory. By 1900 they had built 1200 stationary engines and needed to expand their factory. Instead, they relocated to Charles City, Iowa, where they could set up a larger facility. Their first gasoline-powered traction engine, built specially for drawbar work, followed in 1901. The point about drawbar work is exemplified in the wheels and transmissions of the Hart-Parr tractors. Prior to that the tractors were first and foremost for belt-pulley work. The drive wheels on fragile spoked hubs and spindles were not really up to transmitting torque loads for pulling heavier draft implements.

C.W. Hart (1872–1937) and C.H. Parr (1868–1941) were still engineering students in Madison, Wisconsin, when they made their first two-cylinder engine in 1898. Their first gasoline-powered traction engine followed in 1901.

HART PARR
OIL TRACTORS

The 30 HP Hart-Parr Model Number One. This 17-30 HP tractor, serial number 1205, displayed five 'expansion bulbs' attached to the commercial oil heater radiator modules which comprised the engine's oil cooling system. It was sold to a farmer in Clear Lake, Iowa. Oil as the cooling medium instead of water had the great advantage of not freezing in the winter. Hart-Parr's Model One, serial number 1207, is in the Smithsonian Museum in Washington.*

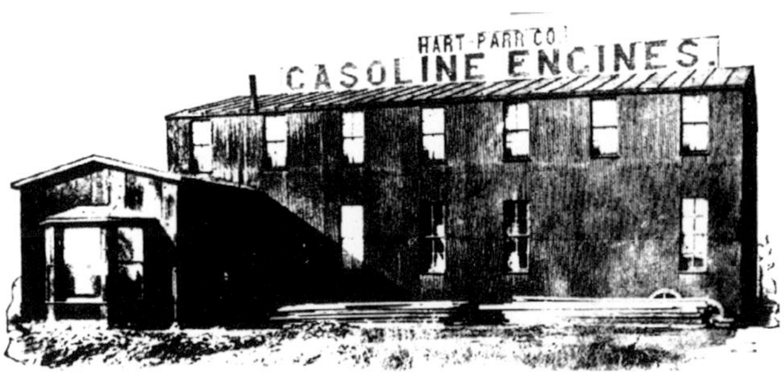

Hart-Parr's Charles City facility became the world's first factory dedicated solely to tractor production (Janssen, 1996). This company was first to use the words 'tractor' and 'farm tractor' in advertising. (Hart-Parr illustrations courtesy White Farm Equipment Company, 1976)

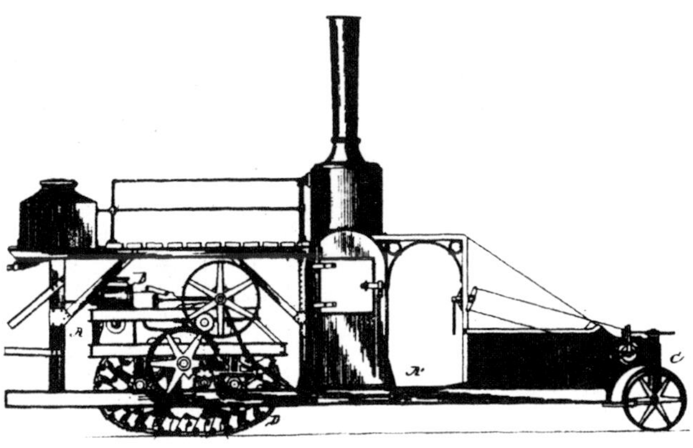

US Patent number 425,600, was issued on 15 April 1890 to Geo. H. Edwards of Chicago. Edwards was first to use the word 'tractor' on a farm machinery patent.

The Charter Gas Engine Company of Sterling, Illinois, was commissioned to build the world's first I-C tractor, which they sold in 1889. They installed their own 20 HP horizontal single-cylinder engine into a Rumely steam traction engine chassis. Six of them were used to belt-drive threshing machinery in South Dakota. (Illustration courtesy Terence Buckaloo, Sterling-Rock Falls Historical Society and Museum, Sterling, Illinois)

Hart-Parr may have capitalised on the word but they did not build the first internal-combustion (I-C) tractor. The Charter Gas Engine Company of Sterling, Illinois, built the first internal combustion traction engine in 1888/89. This US engine maker was commissioned to build six machines for use on large ranches in South Dakota, 1889–92. The Charter 'tractors' resembled contemporary steam engines. The power-to-weight ratio was not much different, since the Charter was essentially a steam traction engine chassis with their single-cylinder gasoline engine levered into it and, while it could move itself around, its prime use was for belt-drive work, powering threshers.

What About Steam? Steam Engines Were Heavy, and Hungry

Steam traction engines weighed anywhere from 700 to 1000 pounds per HP, using historical units, or 450 to

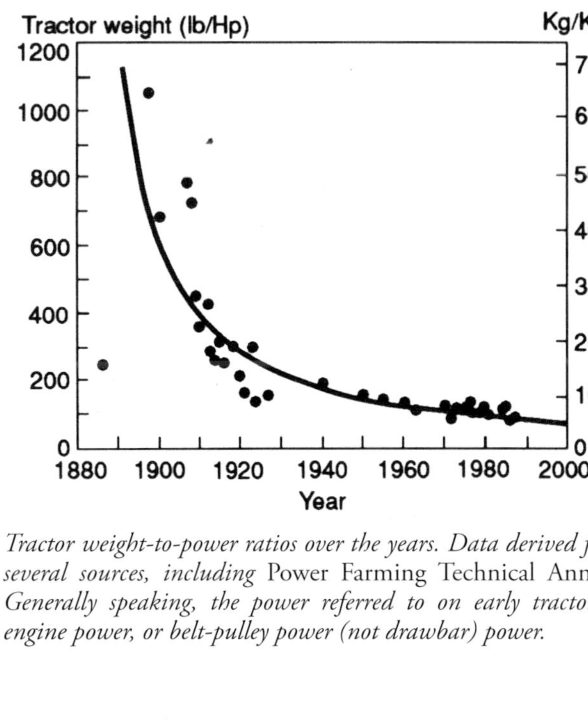

Tractor weight-to-power ratios over the years. Data derived from several sources, including Power Farming Technical Annual. *Generally speaking, the power referred to on early tractors is engine power, or belt-pulley power (not drawbar) power.*

630 kg/kW. A contributor to *Australian Agriculturalist* in 1906 summed it up well: 'there is no better and more reliable type of power known than steam, were the manufacturers able to make an engine light in weight to work on cultivated fields; but that is an impossibility, so that for an all-round farm tractor, steam is out of the question'. The graph on the previous page illustrates the trend in power-to-weight ratio for tractors over the past century as they took over from steam traction engines.

In Britain, steam power is claimed to have played an important role in agriculture, despite the excessive weight and very high cost of steam traction engines. There were 8000 steam engines at work on British farms as far back as 1850 — mostly stationaries at that time (Spence, 1960).

Australians were slow to adopt steam engines in production agriculture. In 1861, an imported steam engine cost £6,000, which put it well beyond the reach of most farmers. There were around a dozen manufacturers who built portable steam engines in the eastern colonies, but they each built only a few engines, and most were deployed in stationary mode for gold-mining. About nine local companies ended up making steam traction engines, but there were fewer than 100 traction engines ever made in Australia. Cliff & Bunting of North Melbourne were the leading makers. Steam traction engines were not only cumbersome, but a steam engine of say eight nominal horsepower (about 25 brake HP, or 19 kW) driving a chaffcutter would consume up to two tonnes of wood a day and 1700 gallons (about 8000 litres) of water — and it had to be clean water. 'If you can't drink it, don't put it in the boiler' was a tested and true saying. In contrast, the compact, one-man operated I-C engines had few of the hassles of lengthy startup time, huge water and fuel demand, fire risk, or the boiler certificate that was needed to drive a steam outfit.

To sum up, steam can hardly be said to have significantly influenced production agriculture in Australia. In 1903, when the first I-C tractor arrived here from England, there was possibly one steam traction engine to several thousand farms and those few were mostly contractor-owned. Today there are roughly two to four tractors on every farm in the country

The First Tractors to Arrive in Australia

An Ivel from England has the distinction of being the first internal combustion engined tractor to come to Australia. Dan Albone of Biggleswade was the inventor of the three-wheeled Ivel and the connection was that he had a brother living in New South Wales. One of his tractors — a special low-profile Ivel — was shipped out to a Tasmanian apple orchardist in 1903 (*Implement and Machinery Review*, 2 January 1904). Several others were shipped to other colonies soon after. An Ivel was sent to New Zealand in 1904 and 17 Ivels were also shipped to South Africa around that time. Norm McKenzie of Cumnock, NSW, owns 40 vintage tractors, including one of the earliest of five Ivels that he has identified that were sent to Australia. The Tasmanian one is believed to be still on the island but buried in a spoil pit! Norm McKenzie's Ivel — serial number 269 — has been restored and is regularly run at vintage rallies. It is one of the world's oldest still-operable tractors. The very first Ivel tractor was donated to London's Victoria and Albert Museum in 1920. The Ivel Agricultural Motor was Britain's first tractor designed specifically to replace the draughthorse — a distinction that could not be given the steam engine.

Ivel tractor designer Dan Albone on his machine. The name Ivel was borrowed from the river near the designer's home town. The Ivel Agricultural Motor was Britain's first tractor designed to replace the horse for drawbar work.

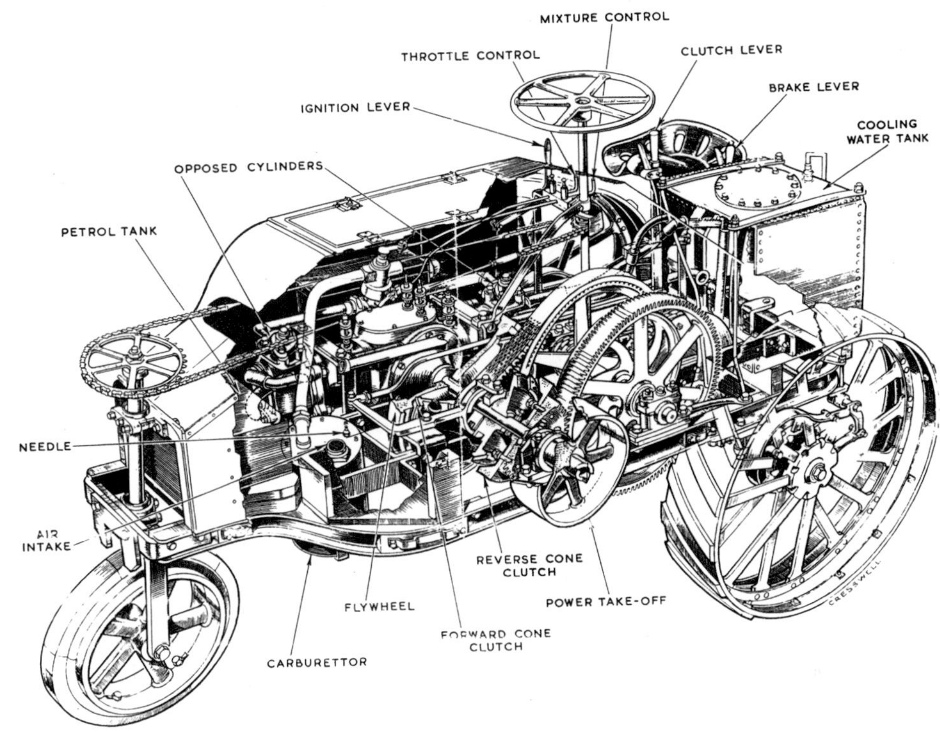

The Ivel Agricultural Motor, 1902. The Ivel tractor was nimble — it weighed 1.85 tons all up, and was powered by a horizontally opposed twin-cylinder four-stroke engine of 24 belt HP at 800 rpm. A twin-cone clutch gave forward or reverse motion to the single-speed transmission. Price of the gasoline-driven tractor was £300 in England in 1905. This outstanding illustration was published in 1954 in the now-defunct British magazine, Farm Mechanisation.

The R. Hornsby & Sons, Spittelgate Works, Grantham, England, were making tractors in England before Dan Albone developed the Ivel. Hornsby started making tractors in 1896, whereas Ivel tractors were first made in 1902. The first of several Hornsby tractors sold in Australia did not reach these shores until 1907 (R. Hooley, personal communication, 1988).

The world's oldest surviving tractor is the Mechwart. Patented in 1894 and constructed by Andreas Mechwart in Hungary, there is an 1896 12 HP model in the Deutsches Landwirtschafts Museum (DLM) in Stuttgart-Hohenheim, Germany. It has not been fully restored to running condition however. Website: www.uni-hohenheim.de

The First Australian-built Tractor

A.H. McDonald & Co. built Australia's first locally made tractor in 1908. Compared with steam engines — or the Hornsby — McDonald's 20 HP was a mere lightweight at 4.5 tons. Its power-to-weight ratio was 504 lb/HP (307 kg/kW), half that of the Hornsby-Ackroyd.

Up to the period of the Great War of 1914–18, the most popular-selling tractors in Australia were the US-built International Moguls and Titans. This was before the Fordson stormed the market. Other makes

Norm McKenzie of New South Wales owns and displays this Ivel at rallies. It is one of the world's oldest operable tractors.

Several of the earliest tractors built by R. Hornsby & Sons failed to sell in England and were accordingly exported to Australia. F.S. Falkiner, Conargo, NSW, acquired his Hornsby-Ackroyd in April 1907 and used it mainly for stationary belt-drive work. This unit was later badly vandalised but is now back in England, where it has been restored. Among the oldest still-operable tractors in the world, the nine-ton Hornsby is powered by a single-cylinder (14.5-inch bore) 20 HP Hornsby-Ackroyd hot-bulb oil engine. (Photos supplied by Ray Hooley, North Hykeham, Lincolnshire)

were imported before the war — including Hart-Parr, Case, Allis-Chalmers, Parrett, Wallis, as well as IHC. But the Australian market was limited. Before World War I horses were still the main source of power for paddock work.

Horse Power

The farm horse population peaked at around 2.5 million in 1919 — or one horse for every two people in the country — and this fell away gradually as tractors were adopted on farms. However, tractors did not really make their major impact until after World War II.

In the 1920s, with horses still the primary source of farm power in Australia, the comparative costs of operating horses, as against tractors, was a hotly debated subject in the rural press. Tractor sales slowly but steadily increased during that decade. In fact, the straight-out economics were generally not in favour of the tractor at first. Then along came Henry Ford's tractors and that brought down capital costs dramatically. A 19 HP Fordson in 1925 at £180 was no more expensive than six good draughthorses, but it could outperform a dozen horses in sustained work.

Several local builders were also on the scene offering tractors in the 1920s: apart from McDonald, there were the Jelbarts, Ronaldson Bros and Tippett, Caldwell-Vale, Sunshine, and Howard, but they were no match in price or marketing for Henry Ford's tractors. In 1928, for example, Ronaldson-Tippett charged about £525 for a tractor of similar power to the Fordson.

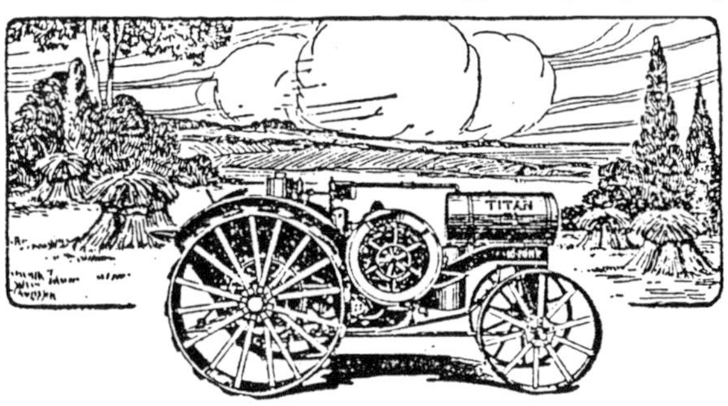

The imported International Harvester Titan 10-20 was one of the leading tractors in the early days of Australian tractor farming (from International Harvester Australia [IHA] advertisement in Land, Transport and Sea Trade, *1919).*

The Frameless Tractor: A New Pattern Emerges

The 1917 Fordson, the Wallis, and similar North American designs with their frameless chassis, established a pattern of compact and relatively light-weight machines, operating at around 200 lb/HP (122kg/kW). The earliest

McDonald's first Imperial tractor, built in 1908, was sold in 1909.

local tractor designs on the other hand, with single- or two-cylinder slow-speed engines, were little more than self-propelling stationary engines for belt work. They were a dying breed by the mid-20s, their days numbered by the higher speed imported multi-cylinder engines.

The locally built Jelbart and McDonald tractors, still with single- or two-cylinder engines, persevered through the Depression, but sales were low. Ronaldson-Tippett, Sunshine and Howard, on the other hand, initially used imported multi-cylinder engines in their production-line tractors.

Henry Ford's ubiquitous Fordson was the world's largest selling tractor model. Fordsons were made in Detroit, Cork and Dagenham, with slight design variations. Almost one million Fordsons were built before the Great Depression, with peak production for a single year of 100 000 units in 1923.

Frank Bottrill built Australia's physically largest tractor. His ultimate machine was the 60 HP, 45-ton (45 kW, 46-tonne) brute named Big Lizzie. Big Lizzie came complete with a blackmithy and anvil on its foredeck. It was used for land-clearing work, opening up

Marlene Quick stands beside Big Lizzie, Frank Bottrill's 45-ton land-clearing tractor. It was built in 1915 and is displayed near Mildura.

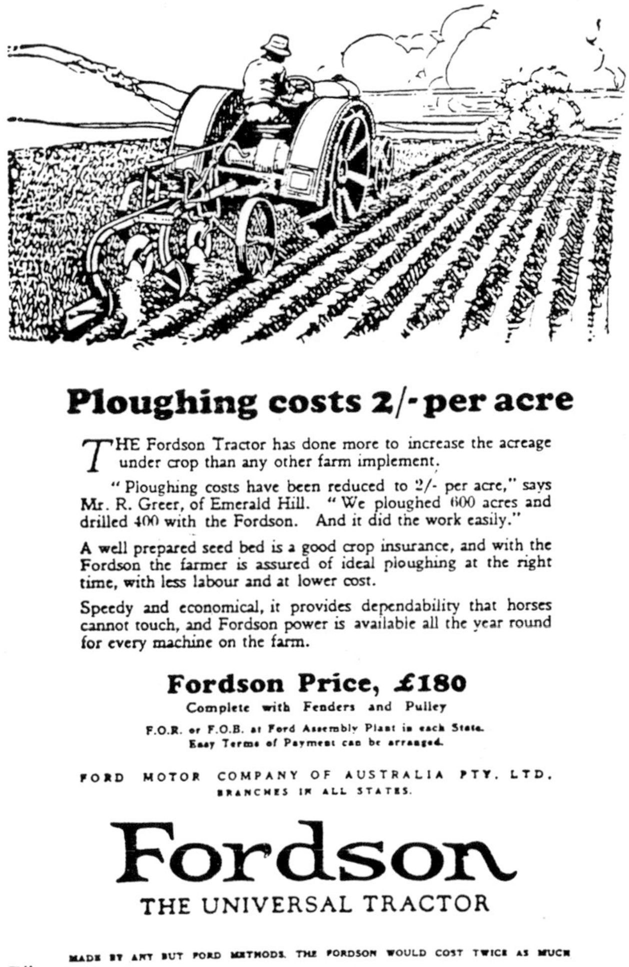

1926 Fordson press release in Australia. Note that the price was just £180 for the world's biggest selling tractor.

farm country in Victoria's north-west. Bottrill, a rugged individualist with a determination to make a firm track, patented his 'Dreadnaught' wheels and used them to give Big Lizzie flotation for contract farm work between 1915 and 1928. The old tractor is now displayed in the town park at Red Cliffs, Victoria.

Early locally built tractors came from the Ballarat works of the Jelbarts and Ronaldson-Tippett, and elsewhere from the McDonalds and Sunshine (Melbourne), Howards (Northmead, NSW), Caldwell-Vale (Auburn, NSW) and Deutscher (Hamilton, Victoria).

During World War II, practically all local tractor production in Australia was suspended and not resumed until 1947. *Power Farming* continued to be issued but its *Technical Annual* was suspended. Immediately after the was, tractor demand exceeded supply. Well into the 1950s many kinds of makes and models flooded in from all over the world, making stiff competition for local manufacturers like Chamberlain, IHA, Kelly & Lewis, Howard, and the original McDonald Company. Other fledgling makers had dropped out by World War II. The best year ever for tractor sales in Australia was 1964, when 23 422 units were sold. A new breed of short-line

Many Others Have Been Relieved by Dr. P. F. E.

'Many Others Have Been Relieved by Dr. P.F.E.' This was the caption for the above cartoon, reproduced from Farm Implement News *(US) by* Power Farming *on 20 December 1930.*

tractor constructors emerged in the 1970s. Some of these even survived the rural downturn of the early 1980s.

Over the years, 25 companies have attempted to manufacture tractors locally. Now there are no tractor builders left and only a handful of constructors building self-propelled machinery in Australia. Altogether, they custom-build fewer than 100 machines a year.

Veteran Australian Tractor Dealerships

The photo collection of Pen Richards, formerly with Howard Rotavator, yielded the views on the right from past Howard tractor dealer/distributors. Howard tractors were doing fairly well until early in the 1960s. Despite having changed hands, the famous Howard name survives on several machines as well as the Rotavator.

Summary

Australia's first locally built tractor was sold by A.H. McDonald in 1909. Twenty-five local manufacturers have produced tractors for the Australian agricultural market, plus another nine builders of steam traction engines and an assortment of makers of self-propelled machinery. At their peak in the 1950s, local tractor producers, led by Chamberlain and IHA, managed to reach an output of several thousand units a year and had carved out for themselves a slice of about one-third of the total market in Australia. But after the 1964 boom sales year, the market began to slide. Since then tractor production has been highly variable, with the fortunes of the farm machinery manufacturers being locked in with the financial wellbeing of Australian farmers. In the face of a shrinking market and heavy-handed

Sam Parkin, the Shepparton agent for Howard and other brands.

The Howard Rotary Hoe agent at Donnybrook, WA.

M. Wilkinson, rotary hoe salesman for Hassett's Victoria, supplied this picture of his 1930s roadster and trailer.

George Rees had an agency for Howard in Mildura.

importations, unprotected local tractor manufacture ceased. The exception to this bleak scenario is the compact equipment business, where Australian manufacturers have set the pace in mini-loaders and garden tractors since the mid-1980s, plus a few intrepid self-propelled machine makers. These are the settings for the scenes in the chapters which follow.

In 1943 Power Farming *used the illustration of the man with the gun on the cover of the July issue. The magazine continued but their* Technical Annual *was suspended through this period.*

International Harvester Company of Australia supported Power Farming *and the war effort with the advertisement on the left in 1942.*

2 THE TRACTOR VERSUS THE HORSE

When we played football we never had to train much. Imagine how many times you'd go backwards and forwards on an acre of land, if you're talking about a seven-inch furrow (horse-drawn single furrow plough) each time. I wouldn't know how many miles I'd walked. But as far as football went we were very strong and solid.
— *Mallee farmer Cecil Armistead, in* Boots and All, *1984*

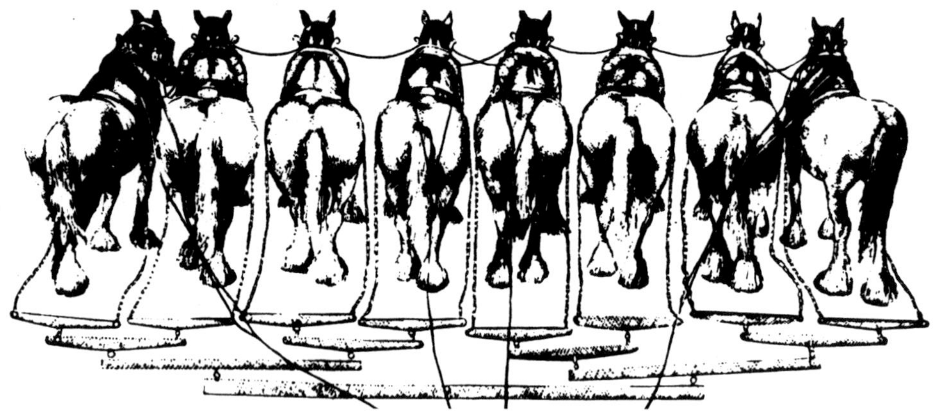

Eight horses abreast. A balanced hitch was essential, and was an art to achieve. The average draughthorse required about five tons of feed a year and then could only work half a day at full load. An eight-horse team would likely be made up of rotations of teams of four horses, each working half a day. Five hours of field work a day was considered a humane limit for horses.

A man walking behind a single-furrow horse plough has to walk 14 miles to cover just one acre. When power farming equipment came along, most operators were delighted to ride rather than walk behind a horse. But was tractor farming in its early days economically justifiable?

Australia's own *Power Farming* magazine (as distinct from the English *Power Farming*) was first published in January 1924. It joined the fray just at a time when the tractor versus the horse issue was being hotly debated. Articles on the debate appeared practically every month in *Power Farming* for the next five years. The issues were presented from both sides, although editorials were of course on the side of the tractor.

Not so supportive, however, were the findings of a conference on Power for Cultivation and Haulage, held at Rothamstead Experiment Station, England, in 1925. 'The tractor will never supersede the horse,' asserted one of the academics at the close of that meeting.

Horse Power

In the days of draught animals, horse power was the method for driving rotating machinery on farms. The horse power used a horse, or pairs of draught animals, either hitched to draught poles and walking around a circular track in a horse works, or restrained on a treadmill or 'endless railway'. These mechanisms were in turn connected to a gearhead which provided power through an output shaft or 'tumbling rod' to the drive pulley for running stationary threshers, corn shellers, feed mills, chaff cutters, wood saws, winnowers, churns, pumps, grindstones, printing presses and so on.

The Costs of Horse Versus Tractor Farming

An article in *Power Farming* in November 1925 observed that, 'A 12 HP tractor does not require half the time and attention spent on it that twelve horses do, if both are to get proper care … no horses should work more than five hours at a stretch without being fed and watered, and it will take them fully one and a half hours to have their hunger satisfied.'

On the other hand: 'the 12 HP tractor costs £500 to £700, whereas a fair horse will cost £25, so twelve at that price is £300. Harness for a horse, collar, hames, winkers and plough chains cost about £3 a set, so we have another £36 to add. To this add the cost of blocks, chains and swingle trees, and many other necessary items which cost, say, £50. That brings the horse team up to £376, so that on that reckoning we have a balance of £224 in favour of the horses. If therefore, we are going to outrule that favour, we have got to find the tractor doing the job better or cheaper than the horses' (*Power*

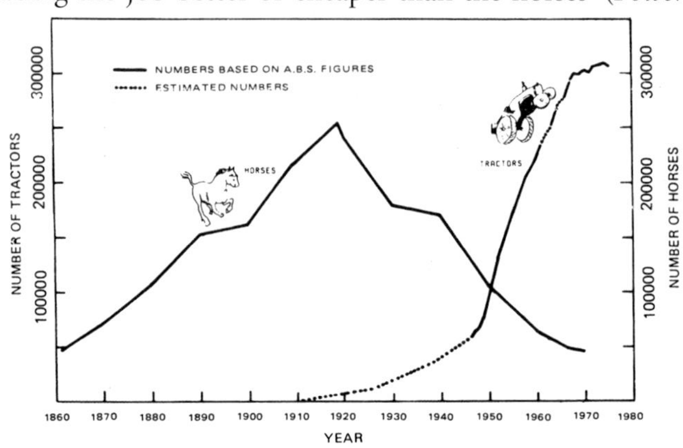

Decline of the horse and rise of tractor numbers on Australian farms, 1860–1975. The same trends occurred in the US, only on a larger order of magnitude. Data derived from Commonwealth Yearbooks *and ABS tractor statistics. Note that horse numbers peaked at around 2.5 million in 1919.*

Farming, November 1925).

The writer was disposed toward the horse team; but then he did not reckon with Henry Ford's cheaper Fordson tractor which, in 1928 was to cost just £195, compared with the local competition at anywhere from £500 to £700. Perhaps the cartoonist got it right in the illustrations, which appeared in *Power Farming* in January 1929 (see page 19).

Power Farming: Original Defender of the Tractor in the Rural Press

The following is dedicated to the editors, contributors and publishers of *Power Farming*, and to Bert Wilson, all individuals who have contributed much to rural Australian life over the years since the magazine was founded in 1924. In the charter year four shillings purchased an annual subscription.

What Is a Horsepower?

The term 'horsepower' was devised in the 18th century by James Watt, who needed an objective standard to rate the power of the newly developed steam power sources. This power unit was derived from the horse's ability to lift a one-ton load up a mine shaft (2240 pounds, then Watt added a safety margin of almost 50% for a total of 3300 pounds) through a lift of 10 feet in one minute, that is, a power output of 33,000 ft.lbs/minute. A strong horse can actually do more work than 'one horsepower' — even as much as five horsepower — for a very short period. An average man, on the other hand, can barely sustain an output of one-tenth of a horsepower without risking a coronary. A

Power Farming *was first published in January 1924. A little self-praise did not go astray in this cartoon — 'The oldest book rack …' — taken from the first edition.*

The rubric of Power Farming*'s first editorial page, January 1924. The 'tractor versus the horse' issue was considered top of the agenda for the lead article, 'Muscle or Machinery?'.*

Bert Wilson's regular columns in Power Farming *commenced in February 1939. At that time the magazine was known as* Farm Mechanics. *That name only lasted for about a year while the publication changed hands. Wartime strictures brought the magazine back to quarterly publication. Wilson's columns have since illumined the pages of* Power Farming *with humour, advice and pithy remarks for Australian tractor and machinery users/owners.*

Two-horse endless rail powering a Fearless stationary thresher and cleaner, 1880.

Scene at Gawler Plains, South Australia, 1876. One-horse endless railway power driving a winnower for cleaning grain dumped from a stripper box. (Photo courtesy Museum of Victoria)

The tractor eventually superseded the horse, but only relatively slowly in Australia. It was the upheaval of World War II that really catapulted Australian farmers off the horse's back. By contrast, the transition in the US was rapid through the 1920s. Horse numbers on Australian farms peaked around 1919, when there was about one horse to every two people in the country for a total horse population of around 2.5 million. The rate of decline in use of horses on farms was slower in Australia. Today there is a ratio of around one horse to 40 people in Australia. Those horses are practically all used for racing or recreation.

A more recent computation of horse costs for feed, etc., versus machinery energy costs in the US, as compared to annual output for draught-horse work, put the cost of the horse at 43 cents per

test at the Research Farm, Werribee, on an eight-horse team of trained Clydesdales in 1949 determined that the peak power per horse was around four horsepower and the average output about two horsepower (W. T. Brown, in *Power Farming*, June 1983).

Formula for tractor vs power of horses: In 1926 the ability of a farm tractor to do the work of horses was calculated conservatively as two-thirds of the drawbar horsepower, and the result, less 1, would give the number of horses on a team which the tractor would successfully replace. For example, a tractor rated at 12 horsepower would replace a team of seven strong farm horses. But unlike the horses, 'the tractor will not tire and can carry on the job day and night, if necessary' (*Power Farming*, March 1926).

Horse fanciers are quick to retort that 'the tractor cannot reproduce itself'. Again, in 1926: 'Tractors being designed for belt work as well as pulling loads are rated with two horsepower numbers. Thus a tractor rated as 15–25 HP means that the engine will give 25 horsepower on belt pulley work and 15 horsepower at the drawbar, or wherever the plough or other implement is attached. The difference between 25 and 15 is roughly accounted for in moving the tractor itself, in driving the lugs into the soil to get a grip, and in the slipping of the wheels' (and drive train losses, it might be added).

*Are you one or the other? Cletrac (Oliver) track tractor advertising (*Power Farming, *January 1924).*

horsepower-hour, versus electric power at 4 cents per horsepower-hour and the average diesel tractor at about 8 cents per horsepower-hour (Chancellor, personal communication, 1980). In the Western world, these are considered the bottom line figures that turn arguments for a return to animal traction into sentimentality.

Nevertheless, the horse retains a unique place in history and in human affections. There can be no counter to Winston Churchill's final word on this issue: 'I have always considered that the substitution of the internal combustion engine for the horse marked a very gloomy milestone in the progress of mankind.'

A day in the life of two of our farmer friends.' Power Farming *ran this cartoon sequence in June 1928.*

Between 1926 and 1929 Power Farming*'s cartoonist dreamed up these mechanical horse contraptions* (above and left). *But it was Henry Ford's tractor, the 19 HP Fordson Model F, that really gave the impetus to cheap tractor operation.*

3 WHY STEAM NEVER MADE IT

There is no better and more reliable type of power known than steam, were the manufacturers able to make an engine light in weight so as to work on cultivated fields; but that is an impossibility, so that for an all-around farm tractor, steam is out of the question.

— Australian Agriculturist, *1906*

Origins of Steam

Mining stimulated the development of steam engines. English engineer, Thomas Savory, patented and built a steam machine in 1698 to pump water from coal mines (Cassell, 1978). Savory's partner, Thomas Newcomen, added a beam and piston to make a practical pump which found widespread use in the mines of Cornwall.

James Watt took out his first steam engine patent in 1769 and began manufacturing engines in 1775 in partnership with British manufacturer, Matthew Boulton (Crowley, 1976). Boulton and Watt engines were quickly adopted by industry. Meanwhile, Watt's colleague, William Murdoch, recognised the potential of steam for transportation purposes and built a working model of a steam locomotive in 1784 (Bonnett, 1965). Richard Trevithick went further and built steam carriages which made a number of journeys at speeds as high as nine mph. At this time, that was considered fast! Trevithick's steam boilers built up a head of 50 psi — about 10 times higher than the pressures at first used by James Watt.

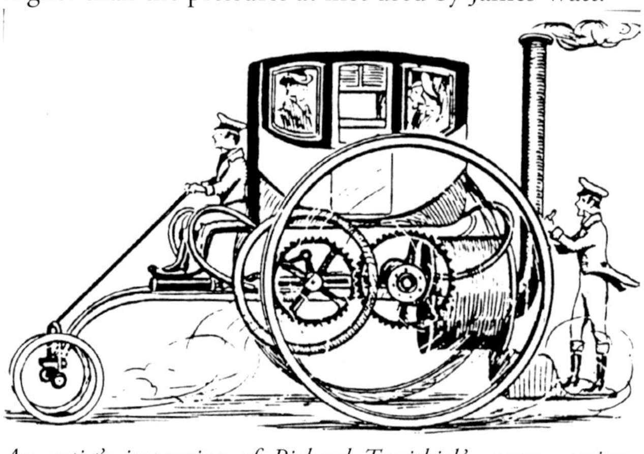

An artist's impression of Richard Trevithick's steam carriage, c. 1803.

The world's first commercially successful steam locomotive was built in 1814 by George Stephenson. He engineered the first railway line to carry passengers and goods. His improved locomotive, the *Rocket*, made its trial trip at 47 mph on the line which he opened on 27 September 1825.

Steam for Ploughing

The idea of using mechanical power for ploughing was recorded as early as 1618. Englishmen David Ramsey and Thomas Wildgosse sought a patent covering 'Newe, apte, or compendious formes or kinde of engines or instruments to ... ploughe grounde without ye horse or oxen'. The application was patent number six and is regarded as the world's first such proposal to harness steam for cultivation (Spence, 1960).

James Watt was clearly aware of the potential and suggested that his steam carriage, patented in 1770, should be able to pull agricultural implements (Bonnett, 1965). The first time steam was actually used in farming was on an estate near Wrexham, North Wales. In 1798 John Wilkinson, a business associate of James Watt, employed one of the inventor's earliest stationary steam engines to drive a threshing drum (Williams, 1977).

Rating Steam Engines

James Watt established the engineering definition of horsepower as the means of objectively rating his steam engines. However, another term emerged in steam engine registration and advertising, namely nominal horsepower (nHP). Nominal horsepower was calculated on the dimensions of the engine (initially area of the piston[s] divided by 10 in England), and later in the US engine speed was factored in. As an example, a 1910 J.I. Case 10 x 10 simple traction engine (10-inch bore) was advertised at 20 nominal HP but could develop 60 brake horsepower (BHP). NHP was not an accurate measure of the engine's true capabilities. Furthermore, not all manufacturers used the same method of rating. The use of the lower rating had much to do with the cost of engine registration.

The First Portable

The world's first portable steam engine was probably the one made by J.R. & A. Ransomes which was publicly displayed in 1841. This engine had its firebox, boiler and engine mounted on a wheeled chassis, and was drawn to the workplace by horses. This machine set a pattern which was rapidly adopted (Williams, 1977).

By 1850 it was estimated that about 8000 portables were at work in Britain (Bonnett, 1965). In the US, portable steam engines were available from several manufacturers by 1850 and were used to drive the popular Pitts and other makers' threshing machines. Still, by 1900 there were only 5000 steam units at farm work on the vast areas of the US.

The steam ploughing rig of Lord Willoughby of Lincolnshire which in 1850 commenced turning the soil 'in an ordinary way by steam'. While one plough was drawn to the centre of the field, the other was being hauled empty to its position on the opposite side of the field by horse. The principles of operation were similar to Heathcote's in the 1830s. (From L'Illustration, *22 June 1850).*

The first working steam-powered cable plough was built and demonstrated in 1836 in England by John Heathcote and Josiah Parkes, four years after it was patented. Some years later that cable ploughing machine was described in one place as 'an experiment of the highest interest'. But then on another occasion the 30-ton monster almost disappeared into a bog, where an observer described it as 'bound to the soil by some sort of mortgage'. Competent authorities conceded that, had the apparatus been tried on better soil than bog, or as a smaller version, it might have been successful (Spence, 1960).

Very High Weight-to-power Ratio

If a cable-ploughing outfit could get bogged, a steam traction engine fared even worse in the early days of steam. Until the weight of the engine could be reduced or the power-to-weight ratio increased, traction machines took second place to cable ploughing outfits. Public acclaim and a Royal Agricultural Society prize of £500 was awarded to John Fowler for a successful steam cable-ploughing demonstration at the Chester agricultural show in 1858 (Bonnett, 1965). Fowler subsequently established a firm which became the major British manufacturer of steam ploughing equipment. In the US, J.W. Fawkes was awarded a medal for his successful steam plough in 1859. Fawkes' machine used a twin-cylinder steam engine, driving cranks geared down 6:1 on each side of the single driving roller, which was six feet in diameter by six feet wide. The British Savory 'boiler-embracing' winding drum was an unusual yet effective machine. When used with another engine on the return cables it was arguably the most efficient cable ploughing system — but also the most expensive.

The Savory cable ploughing rig (above and below) *was a clever design, with the winding drum actually surrounding the boiler. Two Savory engines hauling a reversible plow made an effective but expensive outfit.*

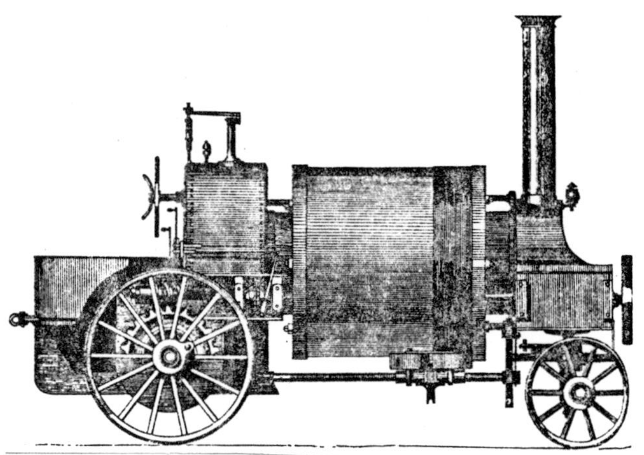

Steam Power to the Wheels: Self-movers and Traction Engines

The Ransomes, Sims and Head Company of Ipswich, exhibited the first self-propelling steam engine for agricultural purposes at an agricultural show. That was in 1842. While their machine was primarily intended as a source of power for threshing, its wheels could be driven through a chain transmission. The firm exhibited another self-moving engine in 1849 known as the Farmer's Engine. This unit was gear-driven, had two forward speeds, and there were springs on the rear axle. Nonetheless it required a horse in shafts at the front for steering (Williams, 1977). That was the law in some states.

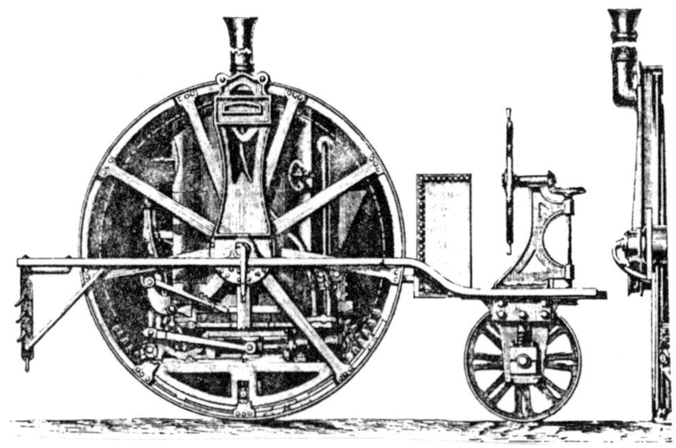

*The Blackburn brothers exhibited this steam traction engine at the Salisbury show in 1857. Inside the eight-foot diameter iron drum was suspended the boiler and a pair of horizontal engines each of 12 horsepower driving an internal spur gear (*The Engineer*, 25 September 1857).*

The practice of using a horse ahead of steam 'traction' engines persisted for many years. There were several reasons:

- There was consideration for other horses on public roads, which were not likely to be frightened by the approach of a steam engine if the machine was preceded by horses.
- The earliest steering gear systems were not considered trustworthy on public roads and it was accordingly safer to guide the machine with horses.
- In some parts it was mandated by road law to have a horse up front.
- There was always the advantage of the extra bit of pull that draughthorses could provide!

A.L. Archambault's 'forty-niner' was one of the earliest commercial portable steam power units in the US in 1849 (Gray, 1975).

A.L. Archambault of Philadelphia built some of the earliest commercial portable steam engines in the US. In 1849 these 'farm engines' — rated in 4, 10 and 30 HP sizes — were also preceded by horses for steerage (Gray, 1975). James Boydell provided an 'endless railway' vehicle for the British army during the Crimean Campaign in 1854 which was claimed to 'draw heavier loads over good roads than any other traction engine, and on soft and bad roads will draw heavy loads where no other traction engine can move itself'.

In 1855 Englishman Peter Halkett patented a 'guideway' system which was a steam-powered bridge or gantry that moved across farmland on metal rails. He installed a small-scale version of the system on his own farm. This was a precursor to modern gantry or controlled traffic farming systems. Also in 1855, Obed Hussey of Baltimore, Maryland, in the US (opponent of famed Cyrus H. McCormick) converted a portable steam engine to a self-propelled 'steam plow' (Gray, 1975).

1880s Russell & Co. outfit on the road. The traction engine was preceded by a horse pair. The thresher is one of Russell's Massillon Cyclone models, the Number 6; the water boy waves from the coal cart (from Russell & Co. catalogue).

A Fowler steam traction engine, with geared final drive to the 14-foot diameter wheels made by John Fowler & Co. (Leeds) Ltd, in 1877. This particular unit was built in Scotland (Clymer, 1959).

James Boydell's 'elephant foot' track-laying engine facilitated traction ploughing as it put down and picked up its 'endless railway' arrangement. In 1856 a Boydell wheel-equipped engine was described as walking 'over the earth like some huge animal, puffing and snorting, and taking along its six plows, with no apparent consciousness of effort' (illustration from Gray, 1975).

C.W. Hoskyns, English farm journalist, lashed out at early draft-ploughing by steam as 'no more to do with the plough than a horse has to do with a spade'. He advocated in 1849 that steam-powered equipment work instead on the rotary principle (Spence, 1960).

Most of the later US steam ploughing engines were self-propelled, whether used for cable-ploughing or for drawbar draft work.

Four years later the English firm, Aveling & Porter, converted a Clayton and Shuttleworth portable engine into a self-propelled machine by fitting a long driving chain between the engine crankshaft and the rear axle (Bonnett, 1985). Based on this experiment, the firm began commercial production of traction engines. Many other makers used similar systems during the following decade and added refinements in gearing, control and steering systems. But they were still very heavy — typically 1000 pounds per HP — and expensive outfits.

Steam in Australia

The first steam engine in Australia was formally set to work by Governor Macquarie in 1815. The stationary power unit was used in a flour mill erected near Darling Harbour by John Dixon, who brought the engine with him from England (Linge, 1969). Dixon's steam-powered mill crushed grain much faster and more regularly than the wind-powered mills that lined the higher points of Sydney at the time. The number of steam engines rose slowly, and by 1850 New South Wales had 75 steam-powered flour mills. Tasmania had about 20 and Victoria had 13, according to state statistical registers. Then came the gold rushes, which accelerated demand for engines.

The importation of steam engines for water pumping and quartz stamping began around 1854, as mechanisation came to the Australian goldfields. Local manufacturers began making steam engines soon after in Victoria, but found difficulty in winning acceptance for their products against the imports. All of the steam locomotives operating on Victorian government railways in 1865 were imports from Britain, despite local makers' protestations that they could match the quality of the British units, even at less cost.

Australian farmers' slowness to adopt steam engines in production agriculture had some justification. The pioneers had to contend with great distances, rough conditions, and unreliable means of transportation for large and heavy items of equipment. The availability of suitable water was also a major factor. The purchase price, up to £6,000 in 1861, put a steam ploughing rig well beyond reach of most farmers. Many settlers had limited capital and what they had was needed for clearing the land and establishing a home and facilities.

Nevertheless, Victoria had 74 steam power units on farms and pastoral properties by 1861, and 329 by 1870 — mostly portables. Two steam ploughs were at work in Victoria by 1862 and New South Wales had its first in 1864. By 1901 there were 333 steam boilers, 289 stationary steam engines, 463 portable steam engines and 53 traction engines on New South Wales farms,

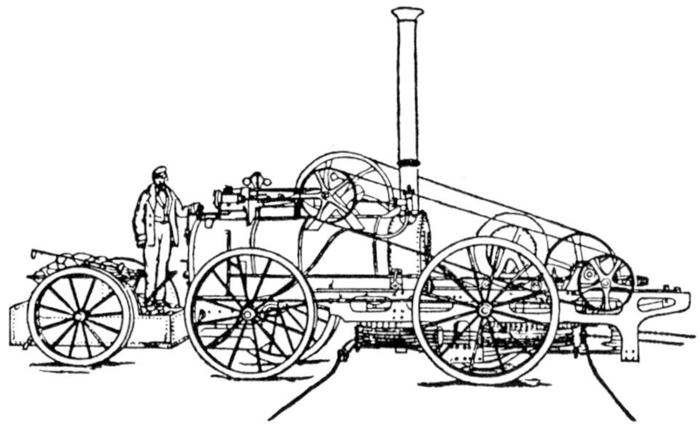

A Fowler cable ploughing outfit from Leeds, England, 1856.

and an additional 872 steam engines used in dairying and pastoral situations (State Statistical Registers and Parliamentary Papers).

In 1888, a Mr O'Shannassy, 'one of the best-known squatters in New South Wales', ordered an 8 HP traction engine with specially built six-furrow plough and cultivator from John Fowler & Co. of Leeds to plough by direct traction (as opposed to cable ploughing). While direct-draft traction was initially unsatisfactory in Britain, where the land was generally too soft to bear the weight of the engine, trials in Victoria and New Zealand had shown that if the ploughing was done in summer and autumn, when the dry soil could better carry the weight of the traction engine, the method could be successful. O'Shannassy maintained that his eight nHP engine could plough six furrows eight inches deep by 10 inches wide on dry ground.

By 1901 there were 23 steam ploughs in use in New South Wales in addition to the 53 traction engines mentioned above. At that time, steam ploughing cost 2 shillings and 2¾ pence per acre to cover five acres per day, with two men to operate the plant (Hayward, 1985). During the 1890s, two large steam ploughing engines supplied by John Fowler & Co. were put into operation on the Melbourne and Metropolitan Board of Works sewage treatment farm at Werribee, where they were used successfully until 1928 when they were replaced by a second-hand pair of Fowler engines. These worked on the farm until replaced by a crawler tractor in 1952. One of these Fowler engines has been restored to operating condition.

The story of two other Fowler engines that were brought to Western Australia in 1889 is worth a mention (Phil Wyndham, personal communication, 1988): Mr T.W. Powell was chairman of the WA Land Company, an establishment set up to build and run the rail line from Beverly to Albany in exchange for land grants. Powell ordered two 16 HP ploughing engines for use on his estates. They left the Leeds Fowler works in England in July and arrived in Albany at the end of October 1889. They were used injudiciously — too deep for the shallow topsoils in the district, bringing up the subsoil. Crops were poor and the estates were abandoned. The engines were bought by the government in 1912, stripped of their ploughing gear and converted to scrub rollers with Bottrill-type drive wheels. This arrangement worked well and halved the costs of land-clearing, but it was hard on the engines, which were sometimes driven over fallen logs up to two feet in diameter. Maintenance costs were excessive. Even so, by the outbreak of World War I over 11 200 acres had been cleared.

Local Manufacture of Steam Power Plants for Agriculture

A number of Australian firms made steam engines of one type or another, starting from at least the mid-1850s. Early manufacturers made stationary engines for mining purposes to meet the demand created by the discovery of gold. Several developed substantial capabilities for producing steam locomotives as rail networks spread across the countryside. Some makers attempted building portable steam engines for an expanding farm sector. Early in the twentieth century, a few firms constructed steam traction engines — just before steam power was overtaken by internal combustion engine technology.

Australian Portable Steam Engines

The manufacture of portable steam engines in Australia was probably started by two Victorian firms during the 1870s, according to the *Australasian Ironmonger* (May 1889). Both continued in production for only a short time. The Fulton Foundry Co., Langlands Foundry Co., or Robinson Bros & Co. were in the engineering business at the time and all could be contenders to have produced the first portable. The same article suggests that the manufacture of portable steam engines began in Victoria about the same time that a duty was first imposed. This was 'approximately 15 years prior to 1889', that is, about 1874. However, relevant customs Acts show that the actual date of the first duty was 1877, when the import duty was '20% ad valorem on ... stationary or portable engines or parts of them'. This was raised to 25% in the mid-1880s. Queensland also imposed a duty, though at a lower rate than in Victoria. Other states allowed free importation of portable steam engines. Local manufacture started shakily and recommenced in 1888, first in Victoria and then Queensland. It was not until 1912 that a New South Wales firm produced a portable steam engine.

A contemporary newspaper report on the Port Phillip Farmers' Society Show held in October 1861, includes a description of a 4 HP steam engine constructed locally by Warhurst & Son (*The Age*, 9 October 1881). The engine was one of 23 steam machines exhibited, of which 19 were portables. Further research is needed to confirm whether Warhurst's engine was one of the portables. If it was, this would almost certainly represent the first portable steam engine made in Australia.

The following Australian makers are known to have built portable steam engines:

Humble and Nicholson, Geelong, Victoria

William Humble came to Victoria from Yorkshire in 1858. After a short period at Fulton's Foundry in Melbourne, Humble moved to Geelong, to join in partnership with Ward Nicholson, who had arrived in Victoria from England in 1857. The partners worked as general founders and engineers, and completed many substantial projects, including the manufacture of hydraulic, refrigeration and agricultural machinery (Sutherland, 1888). This firm made portable steam engines from at least the late 1880s (*Australasian Ironmonger*, May 1889), Surviving engines have single cylinders and are rated in the 2 to 3.5 nominal horsepower range (Burke and Norris, 1978).

The firm was known as Humble & Sons about the turn of the century, at which time it was also believed to have produced a steam truck — in which case, this would have been one of the earliest self-moving road vehicles to be made in Australia, or at least contemporary with John Shearer's steam car at Mannum, South Australia.

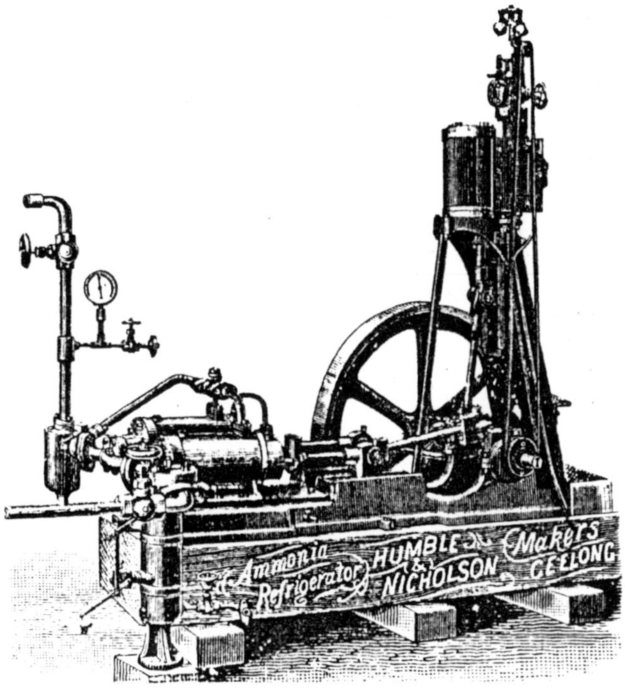

*One of the products of the Humble & Nicholson works at Geelong was this refrigerating plant. The company participated in producing Australia's earliest such machines (*Australian Agriculturalist*, March 1897).*

Messrs Humble & Nicholson's engineering works, Geelong, around the mid-1880s.

David Munro & Co. Ltd, Melbourne, Victoria

David Munro (1844–98) was born in Scotland and migrated to Australia in 1854. He worked in his father's blacksmithing business until it went bankrupt in 1869. The following year he started his own engineering business in Melbourne and quickly built it up to be one of the largest employers of labour in the colony, taking on contracts for railways, bridges and other works, and manufacturing a range of machinery, including Cornish boilers, pumps and windmills. He also sold a wide variety of imported machinery, including portables and traction engines made by the English firms of Marshall Sons & Co., and Ruston & Proctor (*Australasian Ironmonger*, 1887 and October 1888).

In the late 1880s Munro ran into severe financial problems as an outcome of land speculation. He was forced to resign from his engineering works in 1889, but before this he had started making his own portable steam engines. Probably only a few were made. A contemporary account describes them as having a colonial firebox, steam-jacketed cylinder, variable expansion gear, sliding crankshaft bracket and three-way force pumps. Munro exhibited one of his portables at the 1889 Royal Agricultural Society of Victoria Show (*Australasian Ironmonger*, October 1889).

Thompson and Co., Castlemaine, Victoria

David Thompson and his brother James came to Australia from Ireland in the early 1850s. By 1853 they were erecting mining machinery on the Castlemaine goldfields and in 1863 built a steam flour mill there. In 1875 they established a foundry, boilermaking and general engineering works which pioneered hydraulic gold sluicing systems. This equipment was used around the world as well as in Australia. They also manufactured steam bailers, pumping plant, high-speed steam engines, steam turbines, locomotives, steam rollers and other lines (Milner, 1987).

The company commenced making portable steam engines in 1888 (*Australasian Ironmonger*, May 1889). There is no certainty how long they continued making them or how many were made. A surviving example is an 8 HP single-cylinder machine (Burke and Norris, 1978).

Messrs Johnson & Sons, Melbourne, Victoria

John Johnson established Messrs Johnson & Sons after he came to Victoria in 1879 from North Wales. Eighteen months after working at the Phoenix Foundry in Ballarat, he joined Robinson Bros in Melbourne where he became general superintendent before leaving to start his own business. He died in March 1903 (*Australasian Hardware and Machinery*, 1903).

A portable steam engine currently owned by the Melbourne Steam Traction Engine Club Ltd confirms that his company was a local manufacturer of this class of machinery. So far no evidence has turned up to show just when the firm made their portables. Probably only a small number were made. During 1909 the firm manufactured two large steam traction engines. These are referred to in the section on traction engines.

Engine with road gear-built by the Johnson foundry for land clearing at Heytesbury around 1909.

A. Roberts and Sons' United Iron Works, Bendigo, Victoria

The firm was founded by Abraham Roberts (1819–97) who came from Cornwall to Victoria in 1853. After working as a miner and smithy on the Bendigo goldfields, Roberts established a founding and engineering works there in 1860 or 1861 to handle engineering work for Bendigo mines. They manufactured a wide range of mining equipment, including winding, crushing, air-compressing and air-boring machinery. In addition to this mining machinery, the works also made steam engines for riverboats on the Murray, and a range of agricultural implements (Milner, 1987). Presumably the portable steam engine was one of these. It is not known when or how many portable steam engines were made but one survives, as shown.

A. Roberts and Sons' United Iron Works made this portable steam engine at Bendigo during the gold rush. (Photo courtesy P. Milner)

Sidney Williams & Co. Ltd of Sydney, Brisbane, Rockhampton and Townsville

Sidney Williams came to Australia from Britain as a qualified engineer. For a period of time he was involved with droving cattle across the inland stock routes of northern Australia and saw at first hand the difficulty of obtaining water in the outback. In 1897 he founded a business in Rockhampton, which became at the time one of the best known makers of windmills in Australia. His windmill design, the Comet, was extremely simple and reliable, and could be built in a range of sizes with fan diameters from 8 to 35 feet. His windmills were among the biggest in the world. Williams eventually established branches of his business in Townsville, Brisbane and Sydney.

At one stage, Williams also worked on steam pumps. It may have been that which led him to produce portable steam engines, one of which still exists, also known as the Comet.

Toowoomba Foundry and Railway Rolling Stock Manufacturing Co. Ltd (Southern Cross) Toowoomba, Queensland

In 1871, G.W. Griffiths, who migrated from Manchester in England, set up an ironmongery and repair shop in Toowoomba. Shortly afterwards, he was joined by his brother, J.A. Griffiths, an engineering graduate from Manchester University, and the firm became Griffiths Bros and Co. Among the wide range of products which they made were several types of windmill. Later refinements to their windmills became the firm's best-known product. The famous Southern Cross brand

A Southern Cross 7 nHP portable steam engine, made in 1913.

name was used by the company from 1903 onwards.

In 1884 the firm changed its name to Toowoomba Foundry and Railway Rolling Stock Manufacturing Co. Ltd, as a result of its expanding capacity and growing range of products. The company exhibited one of its portable steam engines in 1910, under the Southern Cross logo. The 7 nHP engine was a single-cylinder design (*Australasian Hardware and Machinery,* April 1910).

The Clyde Engineering Co., Granville, NSW

This company developed into one of the state's biggest engineering enterprises, producing locomotives, railway rolling stock, road-rollers, mining and general machinery, as well as manufacturing a wide range of agricultural implements. In 1904 the company began selling its own steam-powered lorry.

With experience in producing steam locomotives, road-rollers, steam lorries and stationary engines, it is not surprising that the firm also released a portable steam engine. This first appeared in 1912 (*Australasian Hardware and Machinery*, April 1910 and August 1904).

*The Clyde steam wagon, made in New South Wales in 1904. (*Australasian Hardware and Machinery, *1904)*

It was noted at the time that the Clyde portable steam engine was the first to have been made in New South Wales. At least one example survives. This is a single-cylinder machine of 8 nHP (Burke and Norris,1978).

Industry commentators reported favourably on locally made portable steam engines of the late 1880s (which probably included assessments of the Humble and Nicholson, the Munro, and the Thompson engines): 'After a close inspection of the whole of these, we find that they will compare favourably with any in this market. The boilers are of the most improved locomotive type, of the best material and efficiently stayed; the cylindrical parts being carefully lagged and neatly covered with sheet iron. In fact all the material used in their construction seems to be of the very best, and the workmanship sound and of superior finish. Economy of fuel has been considered, and the engine cylinders have been taken from a good model, being large in proportion to the h.p., giving abundant area for the expansion of the steam, and the transmission of its full effect to the crank. They are made to work expansively, the steam being cut off at about half stroke, and are fitted with a reversing gear to run in either direction. The wheels, axles, and forecarriage are of most substantial build, and the requirements of the colony have been studied in the construction of the fire-box, so that straw, sawdust or table refuse can be consumed as fuel' (*Australasian Ironmonger*, May 1889).

Other Australian Manufacturers of Portables

A portable engine survives which is believed to have been made by the Melbourne engineering firm of J.C. Renshaw & Co. (R. Butrims and B.C. Roberts, personal communication). John Renshaw (c. 1849–1925) came to Australia from England about 1890. His firm is listed in directories from 1898 until 1915.

Surviving examples show that portables were also made by Overall McCrae in New South Wales, and Littler & Bruce of Rockhampton in Queensland (Butrims and Roberts, personal communication). The Littler & Bruce machine has a two-cylinder engine of about 10 HP.

The Tasmanian engineering firm, Salisbury Foundry Co., is known to have made at least one steam-roller and replacement boilers for portables and traction engines (R. Proctor, personal communication). There is some suggestion that this company may also have made a portable. Several Australian firms made portable boilers. Examples are Atlas Co., Engineers, in Melbourne and Perry Engineering in Gawler (Butrims and Roberts, personal communication).

Steam Traction Engines

The first self-propelled steam engine in Australia was purchased by E.G. Clerk for his property Clerkness, at Bundarra in northern New South Wales. Clerk saw a Boydell-Garrett engine at work on a visit to England and ordered one, for which he paid £500. It was shipped to Australia in 1857. Moving it from Sydney wharf to the Gwydir Valley under its own power was an undertaking. Recognising the problems he would face on the unmade tracks along the way, Clerk made good use of the Boydell 'Self-Laying Endless Railway' mechanism which put down iron 'shoes' under the wheels. Clerk called his machine Megaethon, the Aboriginals called it 'buggy-buggy'. Clerk was only partially successful and bullock teams had to be called in to assist over some of the epic journey. At Clerkness, the 12 nHP steam engine was used primarily to grind grain and cut timber (Clegg, personal communication, 2005; Griffith, 1954).

The few Australian-built steam traction engines were all manufactured in Victoria:

Cliff & Bunting

Cliff & Bunting was a North Melbourne-based enterprise which began making agricultural implements during the 1880s. It specialised in chaff equipment and made a range of chaffcutters which established a reputation for quality.

About the turn of the century, Cliff & Bunting took on the agency for steam traction engines made by the O.S. Kelly Mfg Co., of Springfield, Ohio. Under the direction of Oliver Smith Kelly, this company made engines to exacting standards with seamless boilers which, although expensive, were regarded as the best type available (Norbeck, 1976).

In 1908, Cliff & Bunting began to advertise an Australian-made steam traction engine (*Farmer and Grazier*, January 1908). In one advertisement, the firm boldly proclaimed that 'The Cliff & Bunting is the first Traction Engine manufactured in the Commonwealth'.

Cliff & Bunting described their engine as combining all the best points of the English and American patterns, having heavy boiler plates, centre crank, all-steel gearing, two speeds, through-axle, replaceable bearings, large water tank, and a winding drum on the back axle. In 1910, at the Royal Agricultural Society of Victoria Show, they exhibited the engine driving the firm's latest model of travelling chaffcutter fitted with a patent hay steamer. A show report described it as being '8 HP of their own make ... very neat in appearance, lightness having been attained without sacrifice of strength through the use of best materials only' (*Australasian Hardware and Machinery*, September 1910).

Advertisements for the engine were published between 1910 and 1912, suggesting that it was sold for a period of possibly four years (*Farmer and Grazier*, January 1910 and January 1912). It is not known how many traction engines the firm made, but one example survives.

In 1908 Cliff & Bunting advertised their locally built steam traction engine as 'the first traction engine manufactured in the Commonwealth'. This claim was arguable. (Farmer and Grazier, *January 1908)*

The Quinlivan Four-wheel-drive Steam Traction Engine

In 1908, a Victorian farmer, Thomas Quinlivan, patented one of the most interesting traction engines ever made in Australia. Quinlivan perceived that four-wheel drive would improve traction for the enormously heavy steam engines and so determined to build a design with this as a fundamental feature.

One machine of Quinlivan's design was built in 1910. The Victorian government selected it as one of four later built under contract for clearing and ploughing work on coastal areas of the state. Cowley's Eureka Ironworks began the construction work but proceeded slowly, apparently due to a lack of finished drawings of the machine (Kenyon, 1910). Cowley's works manager, G.F. Wightman, is believed to have made a significant input to the construction (Weikhardt, 1942). When the firm's head, J.E. Cowley, died with the work unfinished, the engine was transferred to Johnson's Tyne Foundry, on the banks of the Yarra at South Melbourne (*Government Gazette*, 1910). It was fitted with an 8 HP, twin-cylinder steam engine (*Government Gazette*, 1909) and was finally ready by October 1910 (*The Leader*, 1 October 1910).

There were several unusual features on Quinlivan's traction engine, apart from its four-wheel-drive. The steam engine was mounted under the boiler rather than above it, as was the case with most British and American machines. All four wheels were enclosed and used as water tanks. This had the great advantage of increasing the amount of water which could be carried, while distributing the extra weight more evenly across the machine, which had power steering and could pull two seven-furrow disc ploughs.

Thomas Quinlivan, centre, standing behind a working model of his four-wheel-drive steam traction engine. The Museum of Victoria has this fully restored model.

Quinlivan's traction engine hauling a road train of wagons loaded with sugarbeets to the factory at Maffra, Victoria, about 1912. (Photos courtesy M.L. Hallett, Museum of Victoria)

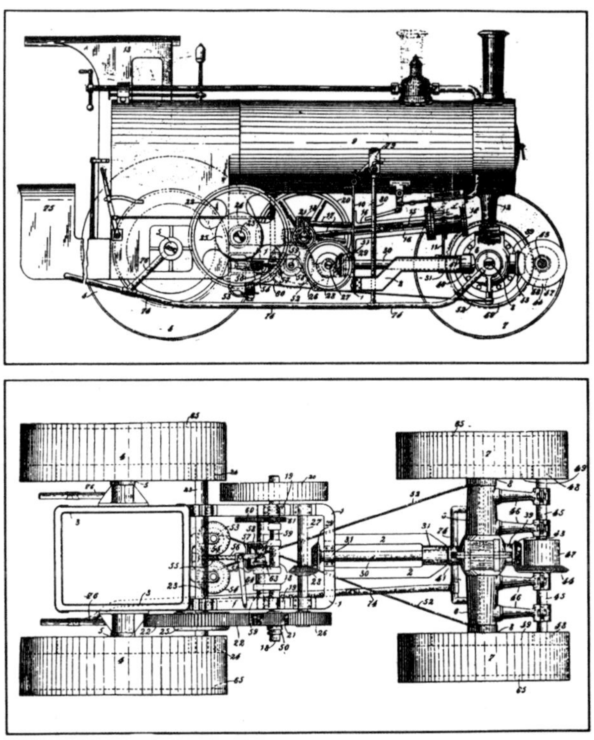

Australian patent number 12781/1908 was awarded to Thomas Quinlivan of Victoria for a four-wheel-drive steam traction engine in 1908.

The engine was used for clearing and ploughing coastal areas, then the Victorian Department of Agriculture moved the engine to a sugarbeet project at Maffra and later to Dookie College. The department disposed of the machine in the 1920s. It was not, however, the world's first four-wheel-driven steam traction. There was a Lansing 4WD steamer in 1898 (Clymer, 1959), and there were at least two 4WD internal combustion tractors in the US by 1910.

Johnson and Sons

As has been mentioned earlier, this firm made portable steam engines and completed the construction of Quinlivan's traction engine. In addition, they were themselves a successful tenderer to produce two steam traction engines for the Victorian government's coastal reclamation program. The firm's engines were very large. A contemporary report described the first of these as being 'far larger than anything that has been either imported or built in the Commonwealth. As an indication of its size, the rear wheels were 8 feet in diameter and 4 feet wide on the tread. This engine was tested under steam in February 1909, closely observed by Mr A.H. Kenyon, Engineer for Agriculture in the Victorian Department of Agriculture, and two officers from the Lands Department of Western Australia' (*The Australasian,* 6 February 1909).

The first Johnson engine left Melbourne in March 1909 for the Heytesbury scheme. Along the way the engine was found to have a number of minor problems: it was not getting enough steam and the gearing was not low enough. The second engine was tested at Werribee in March, where too it was found to need lower gearing. It was sent by rail to Portland. Minor repairs and a change of gear ratio from 23:1 to 42:1 on both engines solved the problems and resulted in the pull needed for rolling scrub and ploughing.

However, in order to work on the soft ground, the tread of the rear wheels had to be widened to 5 feet 8 inches on the Heytesbury machine and 6 feet 8 inches on the Portland. The Johnson engines were rated at 10 HP and able to plough up to 27 furrows at a time. Once the worst of the wet weather was over, both engines performed successfully. A measure of their success can be gauged from the fact that in one year the Heytesbury machine worked a full eight months, while a contractor using an imported machine was only able to work for four months in the same period. Department of Agriculture officers calculated that using the engines in front of the clearing gangs reduced the cost of clearing by 5 shillings per acre. They also calculated that rolling and cross-ploughing with the engines produced a superior result, at less than a fifth of the cost when compared to a settler hand-clearing, then ploughing with two horses (Kenyon, 1910). Unfortunately, the rough working conditions and constant exposure to gritty dust caused a number of components to fail and resulted in excessive wear on both machines. This was another significant constraint on the adoption of steam.

Austral Otis Engineering Co., Melbourne

Austral Otis also successfully tendered to produce two steam traction engines for the Victorian government. The company had its origins in the 1880s as a general engineering business. In the late 1880s it set up a well-equipped works for heavy engineering which covered about four acres and employed over 300 people. It had important agencies for machinery, including Worthington pumps and the Otis Bros & Co. elevators. The company undertook many major contracts for mining and other machinery and was awarded prizes for its steam engines and hoisting equipment at the Centennial Exhibition in 1888 (*Australian Ironmonger*, May 1889). About 1896 the company was engaged to produce four large three-cylinder, triple expansion, stationary steam engines for the Melbourne and Metropolitan Board of Works sewage pumpers. Although these engines were close copies of a British design by Haythorn Davies, their production proved that Austral Otis had substantial expertise and capacity in steam engineering. Austral Otis also produced steamrollers.

The first of the company's steam traction engines for the Victorian government was completed by March 1909. It followed an American pattern, with a vertical boiler and chain drive. Each rear wheel had a tread 4 feet 6 inches wide. By mid-March it was dismantled, transported to Camperdown, reassembled and driven to Heytesbury. The trip was not an easy one — the engine became bogged many times. Getting the engine across Curdie's River proved particularly difficult. A number of problems emerged as soon as the engine was put to work. Some components were inadequately lubricated and a number or remedies were tried. The engine was too heavy for the type of country and could not be used for ploughing. However, the weight of the engine was found to be an advantage for clearing and rolling work, so it was retained exclusively for that purpose.

Johnson & Sons were successful tenderers to the Victorian government to produce two steam traction engines for coastal land reclamation. Rated at 10 nHP, these engines were promoted as a cost-effective technology for land clearing and ploughing. (The Australasian, *6 February 1909)*

An Austral Otis steam traction engine, c. 1909. (Kenyon, 1910)

The company's second traction engine does not seem to have been used by the government. A private company, the Patent Excavator Co., had some modifications made to it and then took it to Yanco in New South Wales for tests. When these proved unsatisfactory the engine was returned to Austral Otis (Kenyon, 1910).

Cowley's Eureka Ironworks Pty Ltd

Some years after the manufacturing activities which had surrounded the Victorian government's land development programs, Cowley's Eureka Ironworks produced what was probably the last steam traction

Moving house. Cowley's 6 HP machine was probably the last steam traction engine manufactured in Australia. It was made at their Eureka Ironworks in 1916. (Photo courtesy M.L. Hallett, Museum of Victoria)

Cowley Steam Traction engine made by Cowley's Eureka Iron Works, Ballarat, around 1913 and one of the last steamers made in Australia. (Photo courtesy Martin Hallett, Museum of Victoria)

engine made in Australia, in 1916. Seeing that the firm had earlier made at least a part of the Quinlivan traction engine, it is not surprising that their unit showed many features reminiscent of that earlier machine, including enclosed wheels which doubled as water tanks. The Museum of Victoria owns this historic steam traction engine and has restored it to working condition.

Australian Steam Machines for Forestry

Imported portable and traction engines played a significant role in logging and milling timber from Australian forests. Although steam power units were a fire hazard in summer, they nevertheless proved popular, because the engines could be fired by offcuts and waste timber — virtually free fuel. Several Australian-built steam machines called 'whims' worked in the forests near the Yarloop district of Western Australia from the late 1890s. They could straddle a log, then lift and transport it to a sawmill.

The idea for the whims was conceived by an employee of Millars Timber Company, Harry Stephen, around 1896. Millars were interested in the idea and gave him every assistance, including the use of the company's workshop.

The original whim, or 'steam transport log carriage' consisted of a wooden framework and platform approximately 25 x 7 feet wide, over four large wheels, almost 10 foot diameter at the front and 11 foot diameter at the back. Power was provided by an 8-inch bore single-cylinder steam engine powered by a vertical boiler. A steam winch and steel rope and chains were used to lift the logs, which were secured under the whim by chains and quick-release hooks. The whim could carry approximately 19 tons at a time. One man was employed to steer the machine and another to look after the boiler. This whim became the first mechanised form of log-carrying in Western Australia.

At least two wooden-framed machines were made.

This prototype steam whim was built locally, using timber for the chassis. It was at Hoffman's Mill with Harry Stephen and one of his sons aboard when this picture was taken on 30 July 1898.

Steam-powered whims were developed in Western Australia for log-hauling. These two views show the steel-framed outfit, which could haul up to 19 tons of timber at a time. A crew of four was needed, two on the whim and two on the ground. It was made by Geo. W. Kelly in Melbourne.

Two units built at Yarloop were designed to haul logs of up to 5 foot diameter and 60 feet long. A steel-framed whim was made in Melbourne by the Geo. W. Kelly Company (later Kelly & Lewis). Its wheels were a mere 9 foot diameter at the rear and 8 foot at the front, but it could still carry 19 tons of timber. The whims worked constantly from the time they were built in the late 1890s until 1914, when the operators went off to the war. They did little or no work after the war, having been superseded by crawler tractors in the meantime.

The whims had their limitations. They were fairly easily bogged in wet ground, and they always had to work close to a water source, since they carried only a limited water supply for the water-hungry boiler. One of the wooden-framed whims was destroyed in a 1930s bushfire. As for the steel-framed machine, it was cut up when the Millar's Hoffman's mill closed in 1964 (P. Wyndham, Rossmoyne, WA, personal communication).

The Decline of Steam Power

In 1888, the world's first internal combustion-engined farm traction machine was assembled in Sterling, Illinois, and sold in 1889 by the Charter Gas Engine Co. Six such tractors (as they were soon to be called) were later commissioned to drive threshing machines for 'bonanza' farms in South Dakota. They were actually self-moving portables for belt work.

Britain's first tractors, Petters and Hornsbys, were built from 1893. Until 1903, fewer than 100 tractors had been built in the US, 60 coming from just two American firms, Huber & Co. of Marion, Ohio, and the Kinnard-Haines Co. of Minneapolis. All these tractors were designed as replacements for steam-powered portables to belt-drive threshing machines and not to work the land.

The lighter weight, one-man-operated internal combustion machines, commonly known as tractors after 1906, had few of the hassles for water, fuel or fire risk. They did not command the constant attention which the early steamers required, and they were faster to start. A steam engine of 6–8 nHP driving a chaffcutter (a common farm use for steam power where it was used in Australia) would consume one to two tons of wood fuel per day and 1430 to 1800 gallons (6500 to 8200 litres) of water, if the hay was also being steamed (Hayward, 1985). Considering that a Furphy water cart carried 150 gallons, then the water-joey was truly kept on his toes to deliver 10 or more tanks a day. And that water had to be clean: 'if you can't drink it, don't put it in the boiler' was a hackneyed but fairly accurate statement about steam outfits. Farmers soon recognised the advantages of internal combustion power over steam or horses.

In the US, sales of tractors jumped to some 4000 units by 1910. But steam was not to be entirely upstaged. The J.I. Case Company, for example, continued on with the production of steam engines, even reaching their peak production in 1912, when they constructed 2322 engines. After that demand fell away and their last steam engine was built in 1920.

Steam never really made it as a force in Australian agriculture. Steam power declined before it even gained any momentum, as is exemplified by New South Wales statistics: in 1901 there were 1677 steam engines in use and just 85 stationary oil engines. Of the 805 steam engines in the farming sector, 463 were portables, 289 were stationary power units and 53 were self-propelled steam traction engines. There were over two million horses on farms. Starting from the time of Federation in 1901, there were no internal combustion engine-powered tractors. By 1911 the number of steam traction engines in New South Wales had increased from 53 to 194, but by then there were 70 oil-powered tractors and 1323 stationary oil engines (Statistical Registers of New South Wales, 1901 and 1911). After 1911 internal combustion completely dominated the engine business.

Without exception, tractor makers in Australia did not dabble with steam. So the steam era passed without a lot of local input, the majority of the steam traction engines having been imports. Nonetheless the enthusiasm of steam buffs for the surviving steam machines has been undiminished with the passage of time.

In summary, only one farm in several thousand across the nation could boast of a steam power unit. Horses then were the primary power source. Today, steam and horses are gone from production agriculture and the average Australian farm has three or more tractors.

[A significant part of this chapter — especially regarding Australian steam-engine makers — was prepared by Martin Hallett, Museum of Victoria. Phil Wyndham of Rossmoyne, WA, provided the information on the steam whims and other useful material.]

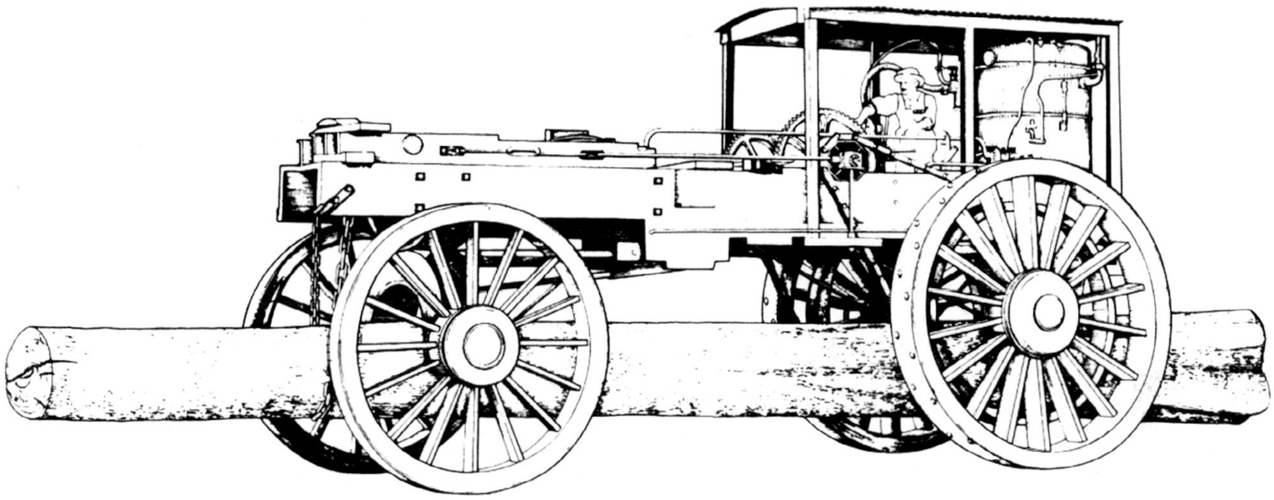

A wooden-framed steam whim built at Yarloop c. 1904–06. The front axle was arched to provide clearance for the logs. Overall length of the machine was 30 feet. (All photos of whims supplied by Phil Wyndham of Rossmoyne, WA, and this drawing is based on one of his photos.)

4 TRACTOR FUELS — CASE TO LITRE

From all the various energies available, the most valuable and the least to be wasted is the energy of the creative mind.
— *Wilhelm Ostwald, 1909 Nobel Laureate*

Early steam engines used solid fuels: wood, lignite, peat, coal, coke, or even crop straw — fuels that were usually obtained locally.

Solid fuels are bulky and inconvenient for mobile equipment such as tractors. Kerosene, on the other hand, was a convenient liquid fuel and it was available for the earliest tractors. The first Australian tractors also operated on imported power kerosene. Power kerosene was a satisfactory fuel to run low compression engines but it did not vaporise readily in a cold engine. Special devices had to be developed to volatilise this fuel. Spark-ignited engines were started from cold on a more volatile fuel such as petrol, benzine or naphtha. Once warmed up, an engine with a compression ratio around 4:1 runs well on power kerosene and supplies smooth power if not heavily loaded, but if used under heavy loads or in engines with compression ratios above 7:1 kerosene burns unevenly and causes engine knock, which is very hard on the engine.

Kerosene, petrol, benzine, naphtha, distillate and oils are all products of distillation from fossil (crude) oil. The first successful oil well was drilled by 'Colonel' Drake in Titusville, Pennsylvania, on 27 August 1859. The early refiners obtained 40–50% kerosene from the crude oil and 4–10% from petrol, which in the early days of refining was a useless and dangerous by-product, so it was burned off at the nearest available dump-site, or even poured into the nearest watercourse! The low return of petrol meant that there was less available and it was more expensive than kerosene.

In 1909, when A.H. McDonald's first tractor was offered for sale, kerosene cost 10¾ pence per gallon (about l.97 cents per litre), compared with petrol at around 1s. 8d. per gallon (3.6c/L). Both fuels were imported, either from the US or from British interests operating in Arabia or in the Far East. Local kerosene derived from shale was primarily lighting kerosene, or 'lamp oil', as distinct from the power kerosene used for engines.

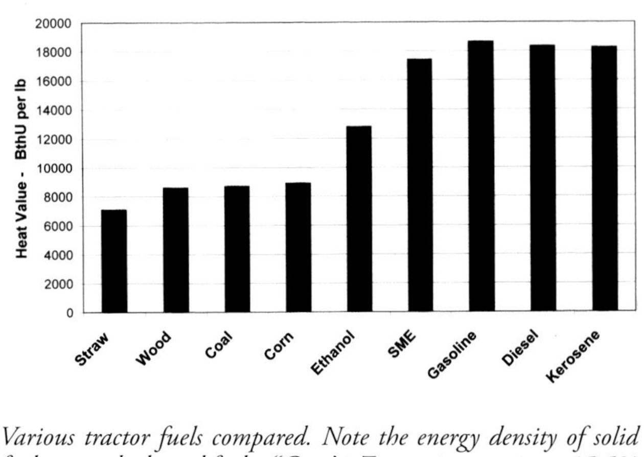

Various tractor fuels compared. Note the energy density of solid fuel versus the liquid fuels. ('Corn' is Zea mais *as grain at 15.5% MC; ethanol is 100% ethyl alcohol; SME = soybean methyl ester.)*

By the end of World War I, the automobile had become a force to be reckoned with. Demand for petrol escalated. Since automotive engines were only designed to run on petrol, up went the price.

Engines and fuels have developed together. Oil refiners changed their techniques for thermally cracking by continuous rather than batch processing. In the process they raised the octane as well as obtaining a much higher petrol yield from the barrel of crude oil. The price differential in favour of kerosene continued but the gap was narrowed. In the US, the price of kerosene was about half that of gasoline. In Australia the price relativity fluctuated with market demand, as seen in the accompanying graph.

Oil Shortage?

As far back as 1919 there was talk of oil shortages! In the period immediately after the Great War the demand for gasoline seemed certain to outstrip crude oil supply. A contemporary headline read: 'Will the earth's oil run out?' The article, which appeared in the local *Land and Sea Transport* (July 1920), continued: 'we are satisfied that close investigation should now proceed, with the object of providing alternative supplies of motor fuels derived from new or supplementary raw materials'. It concluded: 'the importance of the matter (of making Australia independent of outside supplies of liquid fuel by producing liquid fuel from coal) can scarcely be over-estimated'.

The headline of another article declared: 'Oil today is more precious than gold'. That was at a time when petrol sold for about six cents a litre!

The cost-benefit ratio in favour of kerosene continued throughout the era of the spark-ignited tractor engine. Some makers went even further to make it possible for a spark-ignition or Otto-cycle engine to run on even heavier grades of distillate, which were still cheaper fuels. Ronaldson-Tippett in Ballarat, for example, advertised

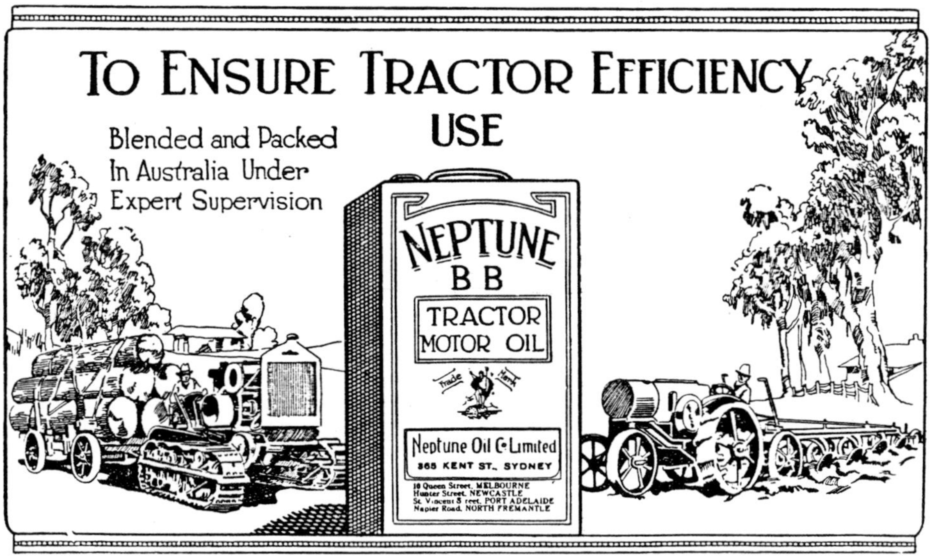

*Neptune engine oil for tractors was an imported petroleum product, blended and packed in Australia. (*The Farm Journal, *January 1922)*

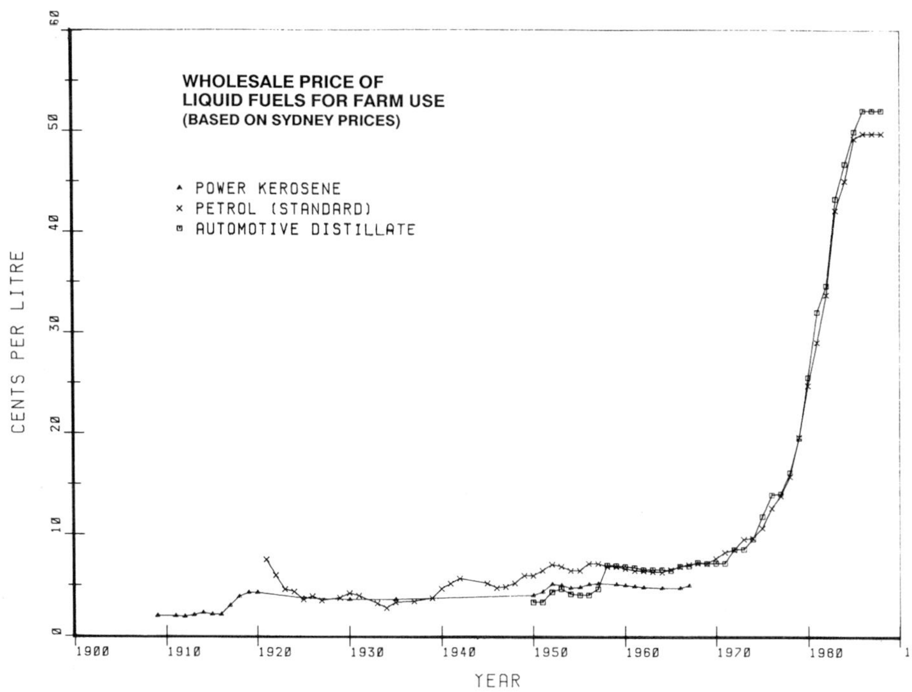

Fuel Price relativity, 1909–90. (Sources of data: various issues of Australasian Ironmonger, *Bureau of Statistics,* Power Farming *articles, manufacturers' sales literature, and the Australian Institute of Petroleum.)*

their Super Drive tractor circa 1926 to operate on 'heavy fuel oil'. Fuel oil was about half the price of kerosene. The company was at pains to point out that this was 'not a diesel' engine. The engine was a simple low compression spark-ignited design.

There is not a large difference in the respective heat value of these fuels. Petrol, kerosene or distillate are each around 18 300 to 19 500 BTU/lb (lower calorific value), but kerosene is slightly more dense than petrol so the heat energy per gallon is about 10% higher. Kerosene was often touted because it was safer to handle than petrol. On the other hand, engines needed to be kept in good condition to burn kerosene satisfactorily. Combustion chamber carbonising was a regular service problem. Injecting water into the manifold was necessary to prevent severe engine knock and slow carbonising when these early engines were operating on kerosene under heavy load.

An early British journal stated that 'at present oil engines leave a lot to be desired in the matter of smell, in fact the effluvium from a single can of paraffin [kerosene] would not only permeate tinned victuals; but would give a flavour to all the water in the Thames and taint the air for five miles around' (cited by Edgington, 1980).

Diesel Engines

Rudolph Diesel's earliest compression-ignition engines were tested on a wide range of fuels — from coal dust to petroleum distillates to vegetable oils. Indeed, at the 1901 Paris Exhibition, a diesel-cycle engine was demonstrated running on peanut oil, which at the time was predicted to become cheaper than coal-derived fuels.

Diesel-cycle engines and fuels suited to compression-ignition developed together. Today's diesels have extremely high-precision fuel injection equipment, designed to operate on tightly specified distillates. The modern fuel distribution system is a marvel of product-handling, especially considering the problems with algae, condensation, dirt and water that can contaminate fuel if it is carelessly handled or stored.

Caterpillar Tractor Company, Peoria, was the first production tractor maker to offer diesel-powered equipment for farm use in North America. That was in 1931, with the release of their Model 65 crawler. The Caterpillar 65 was powered by a four-cylinder four-stroke engine which was started by a two-cylinder pilot motor. In 1935 Caterpillar announced three new diesel tractor models and its 10,400th diesel engine (Gray,

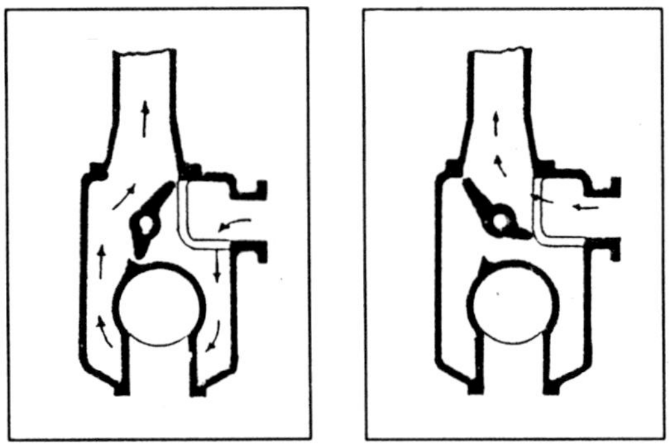

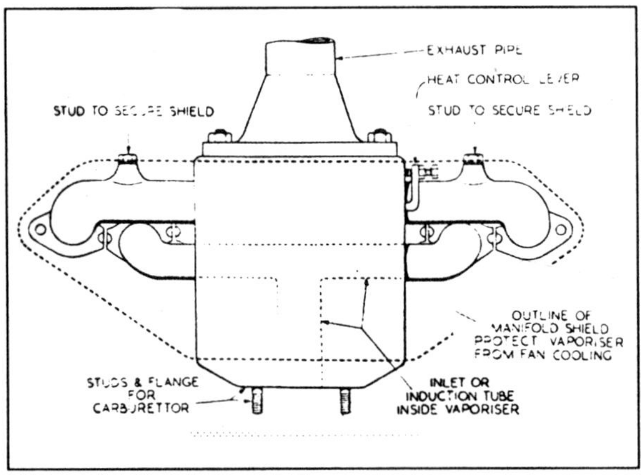

Vaporising manifold for engines running on kerosene. Power kerosene was the major tractor fuel from the earliest days until after World War II. It was generally cheaper than petrol and was acceptable in low compression engines.

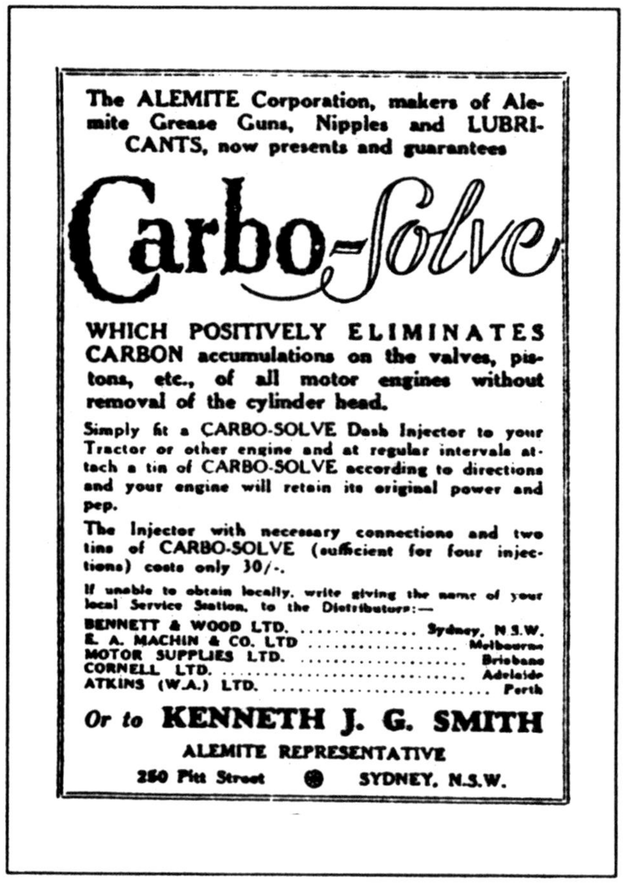

The ALEMITE Corporation, makers of Alemite Grease Guns, Nipples and LUBRICANTS, now presents and guarantees

Carbo-Solve

WHICH POSITIVELY ELIMINATES CARBON accumulations on the valves, pistons, etc., of all motor engines without removal of the cylinder head.

Simply fit a CARBO-SOLVE Dash Injector to your Tractor or other engine and at regular intervals attach a tin of CARBO-SOLVE according to directions and your engine will retain its original power and pep.

The Injector with necessary connections and two tins of CARBO-SOLVE (sufficient for four injections) costs only 30/-.

If unable to obtain locally, write giving the name of your local Service Station, to the Distributors:—

BENNETT & WOOD LTD. .... Sydney, N.S.W.
E. A. MACHIN & CO. LTD .... Melbourne
MOTOR SUPPLIES LTD. .... Brisbane
CORNELL LTD. .... Adelaide
ATKINS (W.A.) LTD. .... Perth

Or to KENNETH J. G. SMITH
ALEMITE REPRESENTATIVE
250 Pitt Street SYDNEY, N.S.W.

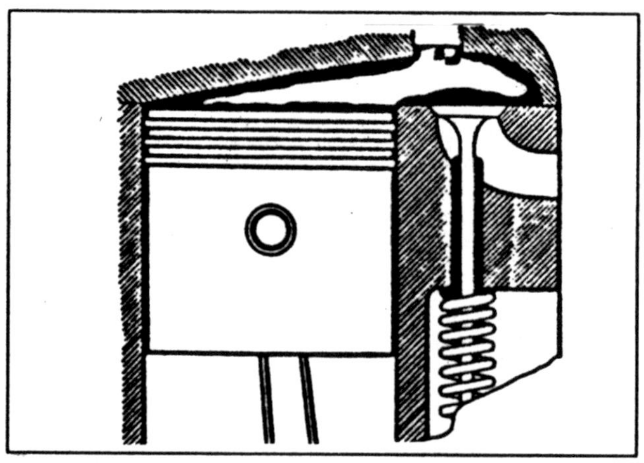

*Carbo-solve advertisement by Alemite (*Power Farming*, 25 August 1930). Engines running on power kerosene had a tendency to regularly 'coke up' and needed fuel additives.*

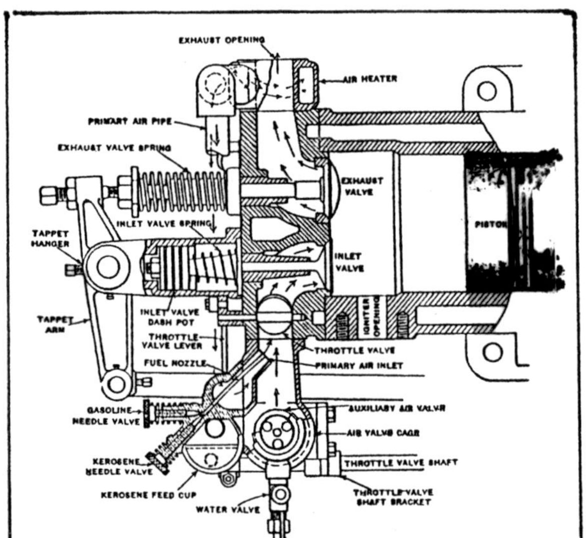

One system for dispensing three liquids into an engine: IH's 1920 stationary oil engine was equipped to meter water as well as petrol and kerosene.

1955). Other companies rapidly followed suit: Oliver, Cleveland, Allis Chalmers — the latter perhaps first with a wheeled diesel-powered tractor in the US. In Europe, Benz-Sendling offered a 'solid-fuelled' diesel-powered tractor in 1922.

The advantages of the diesel engine, in terms of fuel economy and ability to use a cheaper fuel, were not lost on Australian tractor builders. But to develop a farm diesel engine here — that was another story. There were a few makers of stationary diesel engines in Australia before World War II, but these were smaller units, generally unsuitable for tractors. The few low-speed, single-cylinder diesels that did find their way into tractors were soon outpaced by more modern multi-cylinder imported engines.

Hot-bulb Engines

The diesel had to be more ruggedly designed and larger

(Left to right) *John Fell and Co. Ltd, fuel advertisement (*The Farm Journal*, 1 November 1934).*
*Voco advertising by the Vacuum Oil Company (*Power Farming*, 20 April 1935).*
*Cross power kerosene was advertised in 1928 with claims of substantial saving on every acre (*Power Farming*, 25 September 1928).*

*Trident brand power kerosene advertisement from Neptune Oil Co. (*Power Farming*, April 1950).*

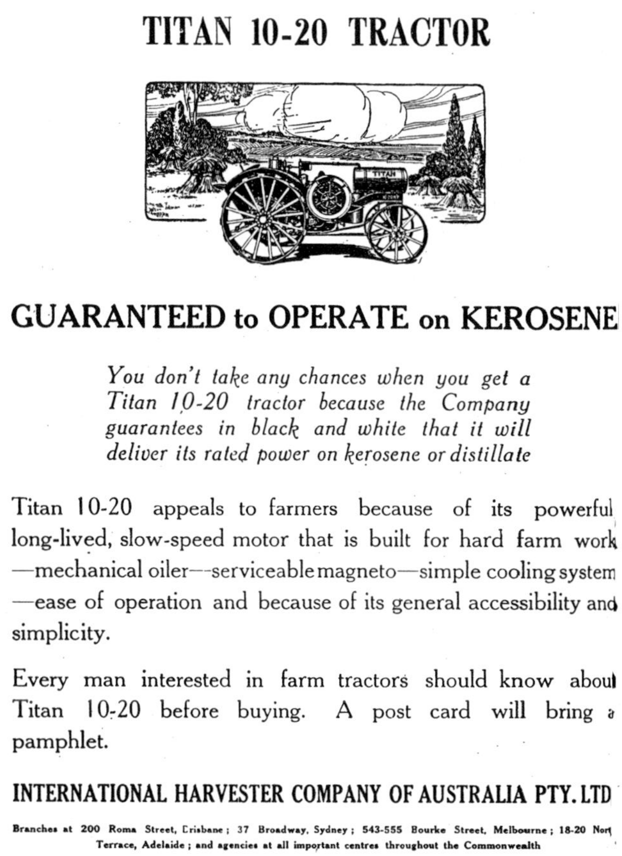

*The Titan 10-20 from International Harvester Company of Australia was guaranteed to operate on kerosene (*Land, Transport and Sea Trade*, 1919).*

Benz diesel tractor 1929. Benz-Sendling built the world's first diesel-powered wheeled tractors in 1922. Several units were brought to Australia. There is at least one in Western Australia and one in South Australia.

Fuel by the case. The wooden protective fuel marketing case usually held two four-gallon fuel tins. It was a convenient size for handling and transport and would fit well on the running board of older motor cars. Farmers often bought their fuel by the case before the 1930s.

than a spark-ignited engine of equal power to cope with the higher compression ratio. Diesels can be difficult to start under cold conditions — requiring special attention to starting methods.

Kelly & Lewis in Australia thought that they had resolved all these problems. They simply adopted the Lanz system, lock, stock and barrel. The trouble was, their tractor was practically out of date by the time they had tooled up and reached the marketplace in 1949. Even then, they were still in trouble, with material quality failures in stressed components. The Lanz engine was a 'hot-bulb' type of semi-diesel, with a modest compression ratio. Hot-bulb engines had an uncooled or unjacketed bulb as portion of the cylinder head. The hot head idea was first patented by Englishman Herbert Ackroyd-Stuart in 1890 and actually preceded Rudolph Diesel's patents.

To start the KL or the Lanz Bulldog engine, which was a two-stroke, the dome or bulb had to be heated by a blow lamp. The fuel was injected into the combustion space by a small 'sprayer' at modest pressure just before the end of the compression cycle, and was ignited by the hot surface. Air only was drawn in through the inlet valve and, once running, compression plus the hot cylinder head heated the induced air sufficiently to keep the engine running. The hot-bulb engine in Jelbart tractors, for example, required a fuel of lower self-ignition temperature than the power kerosene used in spark-ignition engines, so lighting kerosene was preferable, although many hot-bulb engines could run on a range of the heavier fuels, including recycled sump oil. In any case, it was recommended that the sprayer and hot bulb be examined frequently. Carbon deposits had to be removed after every 50 hours of operation.

McDonald produced simple low-speed diesel engines from 1924 onwards, until overtaken by imported high-speed diesel technology in the 1950s. In 1954 Chamberlain modified a 40K kerosene-burning two-cylinder two-stroke engine to operate as a full (solid-fuel) injection diesel.

Do not be confused, incidentally, if you read about 'oil tractors' in early advertisements. These were usually spark-ignited kerosene-burning engines. Before World War I kerosene was known variously as vaporising oil, power kerosene, paraffin, or even as 'distillate'. The use of the term 'oil engine' was confusing — most were spark-ignited, kerosene burners. But there were others, some made by the same company, that were hot-bulb type semi-diesels which had the great advantage of no electrical ignition system.

To overcome problems with engine knock and low-grade fuels, most of the kerosene-fuelled engine makers made provision for water to be introduced into the combustion chamber one way or another when kerosene was being burned under heavy loads. Such engines could consume as much or more water than kerosene. The 1890s US patents of Secor and Higgins (Rumely Company) set the pattern for carburettors to meter three different liquids — water, petrol and kerosene. Those types of carburettors were also used on Australian tractors.

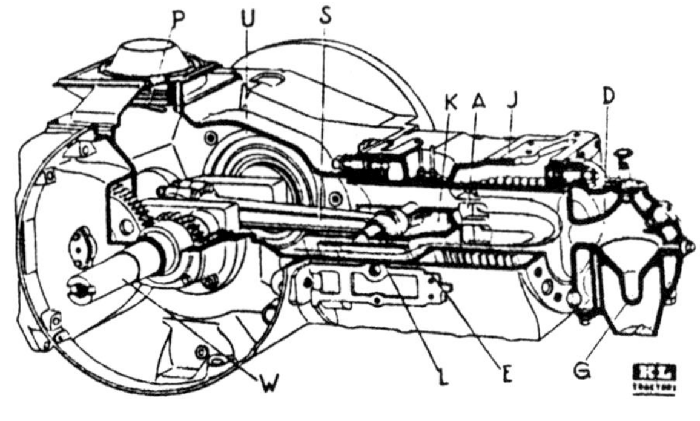

Kelly & Lewis KL Bulldog two-stroke engine schematic.

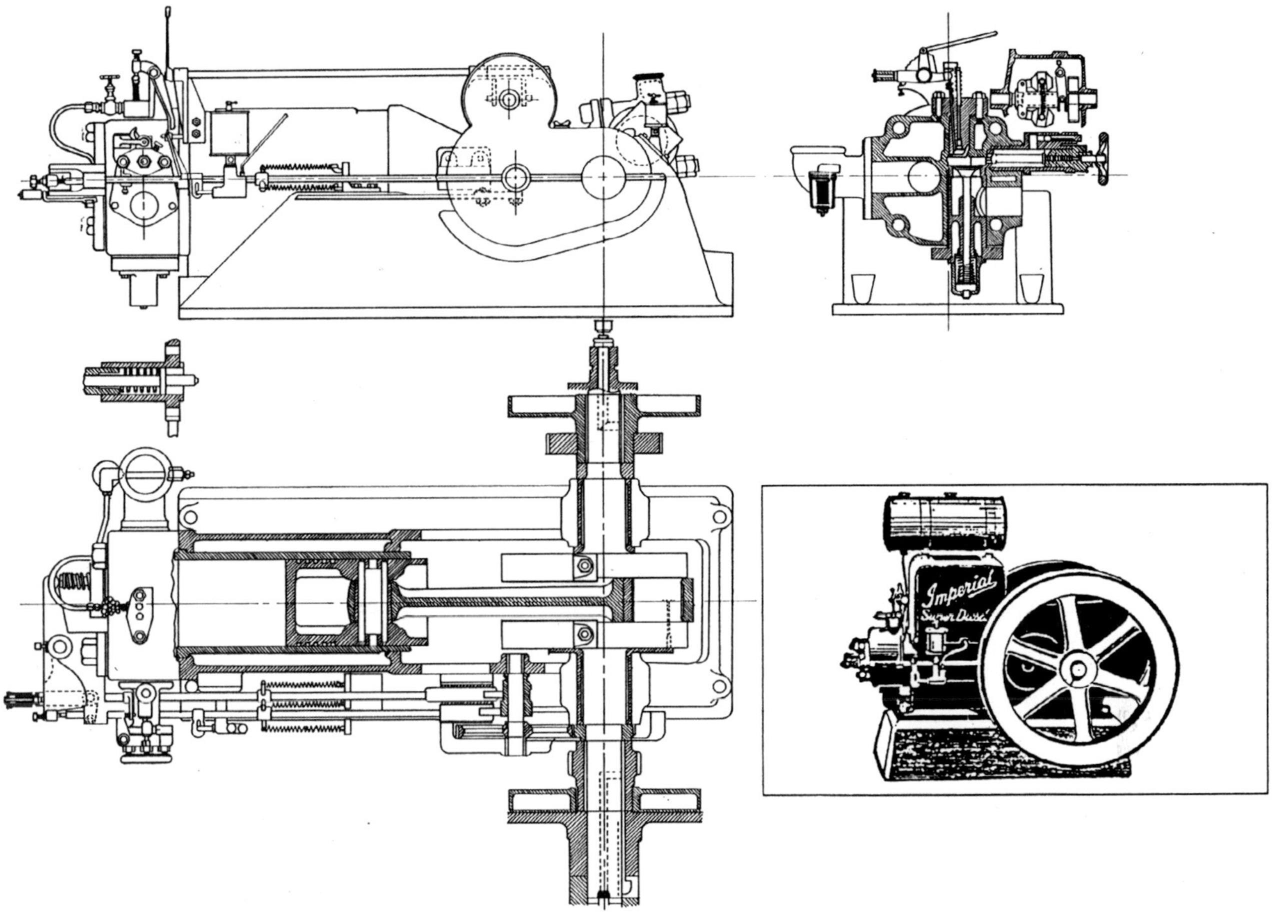

McDonald Super Diesel engine schematic. Inset: illustration from a 1926 company advertisement.

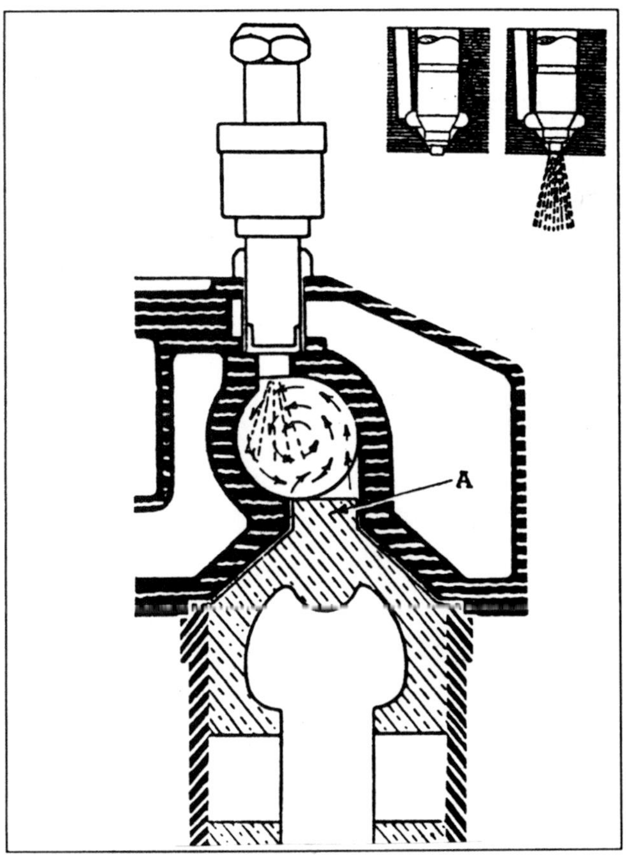

Cooper four-stroke diesel cylinder head schematic.

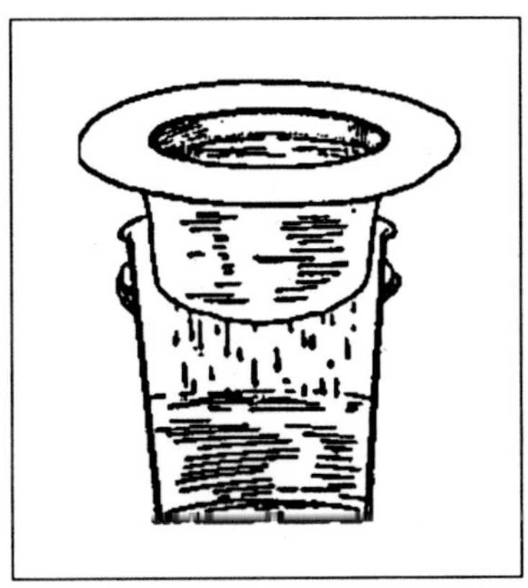

Oil filter. One correspondent suggested that the old felt hat would 'filter lubricating oil perfectly' (Power Farming, *15 August 1924). Don't quote this book as justification for such an oil recycling practice!*

5 PRODUCER GAS: THE CHEAPEST TRACTOR FUEL

World War II was worse for rationing and getting supplies. There was petrol rationing for starters. You'd only get a few gallons and it'd be gone before you turned around. There was an option: a charcoal-burner. They were the most horrible things you could imagine. Oh, they were terrible things! I never tackled them. I'd sooner push the bike. But a lot of people put them on. You got the charcoal and put it into a container and poured water on it — I've forgotten the details. But it was terrible! Half the time you were stuck up on the road with your thing blocked up.

— John Whyte, in Boots and All, *1984*

The cheapest fuel of all was, and under some circumstances still could be, producer gas. It is also known as 'suction', 'water' or 'charcoal' gas. This combustible gas is produced in a container known as a gas generator or gasifier.

Producer-gas generators, which are still around, have been used for heating, lighting and cooking back to the 1600s in Europe. Don't confuse these with biogas generators which generate methane by biological processes. Producer-gas generators, or 'gas producers', generate combustible gases from a variety of carbonaceous solid fuels by pyrolising the fuel in a container over a grate with a restricted air supply, that is to say, in a reducing atmosphere.

The basic reactions that take place in the gasifier are as follows:

- Oxidation of the carbon in the fuel by the oxygen from the air supply, with partial combustion, but in the absence of sufficient oxygen:

$$C + O_2 = CO_2 + \text{Heat}$$

$$\text{and } C + O = CO + \text{Heat}$$

- At higher temperatures (900⁰ C), steam will dissociate in the presence of carbon:

$$\text{Heat} + C + H_2O = CO + H_2$$

At intermediate temperatures (around 700 C) steam reactions also occur:

$$\text{Heat} + CO + H_2O = CO_2 + H_2$$

- In the presence of glowing charcoal (high-temperature carbon) further reduction of CO_2 will follow:

$$\text{Heat} + CO_2 + C = 2CO$$

- Hot carbon and hydrogen produce methane, another burnable gas:

$$C + 2H_2 = CH_4 \text{ (methane)}$$

In a producer-gas generator, the overall efficiency of pyrolytic conversion of solid carbonaceous fuels into combustible gases depends on the temperature of the reaction, the heat content of the ashes, radiation and combustion losses. The composition of the combustible gas output (i.e., excluding the nitrogen) varies also, but it is predominantly CO, H_2 and small amounts of CO_2 and CH_4. Producer gas has about half the calorific value of natural gas.

Producer-gas generators have had a checkered career, being adopted in times of liquid fuel crisis and shortage, and tending to be promptly forgotten or even unceremoniously dumped as soon as supplies are resumed. There is nothing so convenient as a jerrycan full of motor spirit! Consider instead the bulkiness of a load of firewood, coke or charcoal of equivalent energy and the message is clear. One gallon of petrol will do the work of 14 pounds of charcoal (twice as heavy) and it will start the engine promptly as well.

During World War II, petrol, kerosene (and labour) were in short supply or rationed. As liquid fuels were tightly rationed, government departments promoted farm gas producers on the following grounds:

- The fuel energy could be derived from raw materials on the farm: wood, charcoal, straw, dried manure, grain, etc.
- There need be little modification of the engine, although some alterations such as increased compression ratio were desirable.
- Spark-ignited engines run a little cooler on producer gas. There was less contamination of the crankcase oil; oil life was extended and valves and heads coked up less than on kerosene. As a result, operating costs were further reduced.

Howard Auto Cultivators manufactured producer-gas generators for use with their tractors during and just after World War II. A DH22 tractor is seen here working with a gas generator. (J.P. Richards photo)

During the Depression years, some tractor manufacturers also promoted and offered gas producers for these same reasons.

Cost Comparisons for Gas Producers on Tractors

Power Farming ran regular articles on producer-gas generators during the late 1930s and early in World War II. Considerable savings in cost of tractor operation were attainable and farm production costs were reduced accordingly. One Mallee farmer with a Ronaldson-Tippett tractor on heavy work went through 'not more than five to eight bags of charcoal per hour and about 1.5 gallons of water per hour. Initial start is on petrol and the changeover to gas could be made in 2–3 minutes. The charcoal cost one shilling and throopence [sic] per bag and the lubricating oil runs longer' (*Power Farming*, 25 May 1936).

The New South Wales Department of Agriculture conducted an investigation in 1940 and published the results which said, in summary, that the substitution of charcoal gas for tractor liquid fuel offered a saving of 3.5 pence per bushel on a 12-bushel crop (*Agricultural Gazette*, 1 May 1940).

Under full load over a 10-hour day, the comparison was:

Charcoal gas: About five bags at 1s. 3d. per 50-pound bag of charcoal.

Kerosene: 20 gallons at 1s. 3d. per gallon.

The overall costs were £1 5s. 1d versus 6s. 6½d., or a reduction in favour of charcoal of 73.9%. The engine oil lasted an average 100 hours versus 40 hours on kerosene — a cost reduction in favour of charcoal of 65% in oil saved.

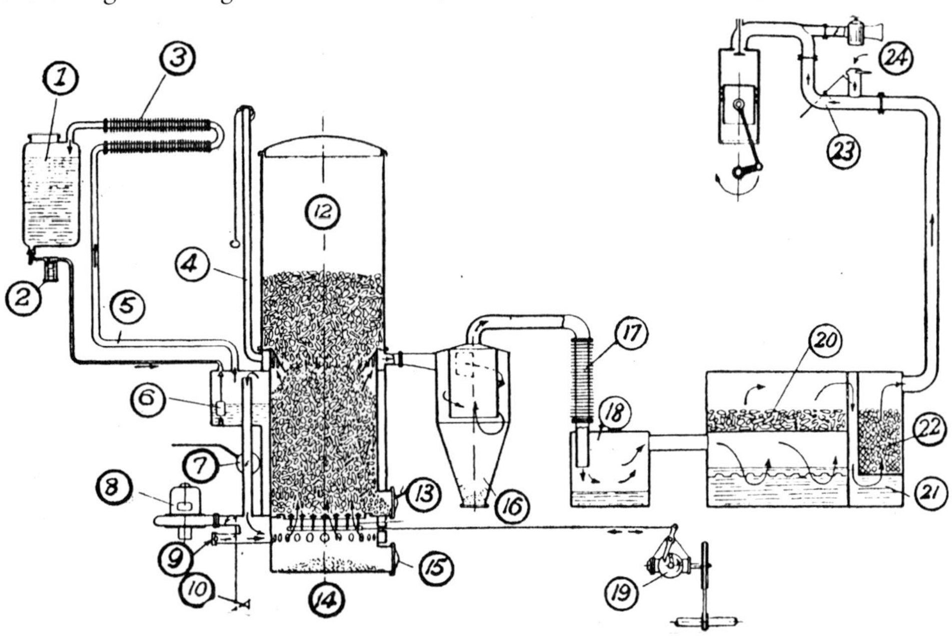

*Schematic layout of a Wisco producer-gas generator unit suitable for tractor operation (*Power Farming, *May 1940). Item 12 is the actual gas generator assembly, which was periodically filled with the solid fuel through the hopper lid. The generated gas had to be 'scrubbed' and cooled, and this required regular replenishment of the water supply in the gas scrubber (21).*

'Eighty Miles to the Bushel'

Mr Bruce Stevens, a former Instructor at the Royal Melbourne Institute of Technology, related how he used wheat as the feedstock for the producer-gas generator mounted on his World War II Durant car.

Charcoal quality was crucial in those days to maintain the gas producer in good working condition, said Stevens, who claimed that 20,000 miles was successfully logged using clean grain only as fuel, and that there was no significant oil wear. The wheat burned cooler than charcoal and there was less clinker. Requiring only monthly cleanouts, instead of the daily chore with charcoal, the grain fuel was filled into a 2/3 bushel container which would sustain 60 miles of knock-free driving. At today's prices for grain, that's cheaper to operate than petrol. This worked out at the equivalent of 38 miles a gallon, compared with the

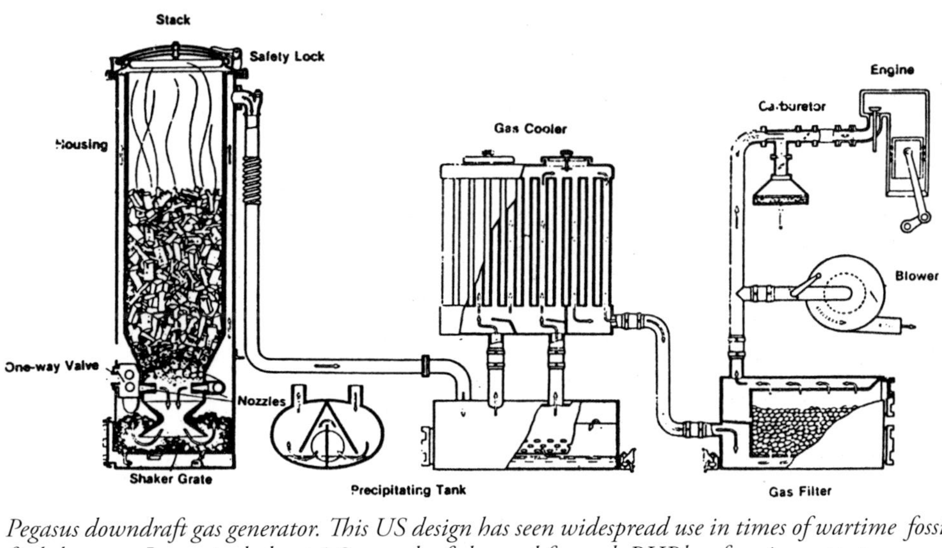

Pegasus downdraft gas generator. This US design has seen widespread use in times of wartime fossil fuel shortage. It required about 1.2 pounds of charcoal for each BHP.hr of engine output.

22 mpg the Durant could manage on petrol. Stevens showed initiative in also devising a unique automatic fuel feeding device to supply the grain by means of an exhaust manifold pressure-sensing meter which controlled the speed of the grainfeed auger, feeding the solid fuel in proportion to engine load. He also devised a producer-gas starting aid, thereby totally avoiding the need for petrol during startup. Stevens mentioned that barley (less awns) worked equally well, and related how on one occasion the grain fuel container sprang a leak. He left a 20-mile trail of bird feed along Melbourne streets!

Gas producers came in for academic assessment again in the period 1979–83 with the fossil fuel crisis. At that time it seemed that burning wheat directly in a producer-gas generator might have been a more efficient way to use the grain than to convert it to ethanol or put in onto a market with low grain prices (*Agricultural Engineering Australia*, 1980). The concept was only evaluated experimentally, since the energetics and economics were unfavourable.

Drawbacks of Charcoal Gas Producers

- Fire risk: 'provided care was exercised, no great danger was presented', said the 1940 Report from New South Wales. Insurance cover was extended to cover producer-gas fitted tractors, without discrimination in some states. Without doubt, though, careless use of producer-gas units did result in many fires in the countryside and they received bad press due to this.
- Engine modification: the compression ratio of the engine needed to be raised, and valves and a gas admission system had to be adapted to the tractor manifold.
- Reduced power: without modification, an engine would deliver about 15% less power, and power was down even with engine modifications.
- Charcoal quality: 'home produced' charcoal was often of variable quality. Dirt and inconsistent charcoal quality could cause excessive clinker formation in the grates, and uneven performance.
- Starting: more difficult and slower than on liquid fuels. It took up to 10 minutes to get the gas flowing from a cold start, although this delay could be considerably shortened by starting on petrol and switching over to producer gas when the gas started to flow.
- Inconvenience and dirty operation: the fuel had to be loaded or recharged frequently during the day and the grates and filters had to be cleaned regularly, for example, every four hours of operation. It was no job to be done while wearing a business suit and tie!
- Bulkiness of the fuel: liquid fuels always have a higher energy density than solid fuels and are accordingly much more convenient to handle and distribute.
- Bulk of the gas generator: the device was an ugly appendage on a tractor and could get in the way.
- Finally and very importantly, suitable fuels, such as timber, had to be available in sufficient quantity.

Some operators had trailer-mounted gas generators which they could switch relatively simply from their motor car to their tractor.

In summary, producer gas was, and still could be, the cheapest fuel … but it is no more convenient now than it was in earlier days.

*Home-made gas plant on the tractor of Mr W.J. Nicholas, via Goornorg, Victoria. His 15-27 John Deere tractor conversion was done cheaply and allowed considerable cost savings. However, visibility was sacrificed (*Power Farming*, 1936).*

*Ronaldson Bros & Tippett Gas Producer. The company marketed these for their tractors during the 1930s. (*Power Farming*, 1935)*

*Oil companies beware! This correspondent 'don't want nothin' for this idea — I giv it ter the korse of sufferin' humanity'. (*Power Farming*, 25 September 1926)*

6 POWER TO THE GROUND

For driving a tractor with cleats attached on a shire road, without the written permission of the Shire Council, Leslie Derrick was fined £20, with costs, at the Temora Police Court, NSW, on May 1. A similar charge against an employee of Derrick brought a fine of £5, with costs. In the latter case, McGregor grips were attached to the vehicle. It was stated that Derrick drove a motor tractor, which had 5 in. or 6 in. spikes on the bearing surface of the back tyres, without having been authorised in writing by the Council.

Defendant admitted having driven the tractor while carting furniture. He also stated that it did not occur to him that would damage the road. He used the tractor on his farm.

— Power Farming, *June 1941*

Steel Wheels

Wheels! There have been almost as many variations in wheel patterns as there have been tractors themselves. On the earliest tractors the drive, or bull, gear was mounted directly onto the spokes and braced (sometimes with spring shock absorbers) to the rims of the steel wheels. Such an arrangement was intended for self-propulsion of the engine, and was not originally designed for heavy pulling or drawbar work.

'Farmers who know the inconvenience and loss of time caused through soil packing on their tractor wheels in wet or sticky soil will appreciate the new McCormick-Deering Tractor Scraper.' Advertisement by IH Australia in Power Farming, *December 1924.*

As traction became more important for pulling from the drawbar, drive was directed through the axle via spokes to the rims. The practice of rolling the rims into a steel channel section, then bolting cleats, grips, or lugs onto the rim was quickly adopted as a sensible way of removing or replacing worn traction aids. The big problem was that there was no ideal lug shape for steel wheels that would work best under all soil conditions and types of country. In the 1920s, Ronaldson-Tippett offered more than 20 combinations of wheel and lug options for their Super Drive tractors, as the reproduction from their 1930 brochure on the next page shows. Among these options were 'road lugs' to allow the tractor to traverse public roads without incurring the displeasure of the local authorities.

While road lugs overcame the restrictive laws prohibiting cleats and public roads, they offered only nominal traction for field work, so they were not too popular. Few operators

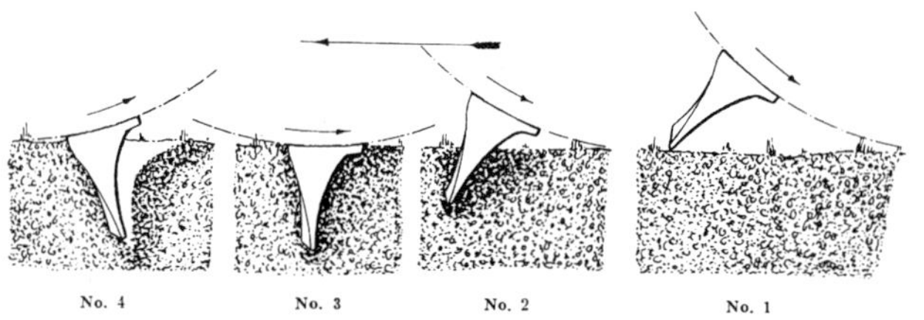

How a lug gets grip on the ground. The lug must compact some soil ('slip') in order to obtain the reaction to propel the vehicle forward. This applies equally to any type of traction lug — whether steel or rubber (from McGregor advertising, 1932).

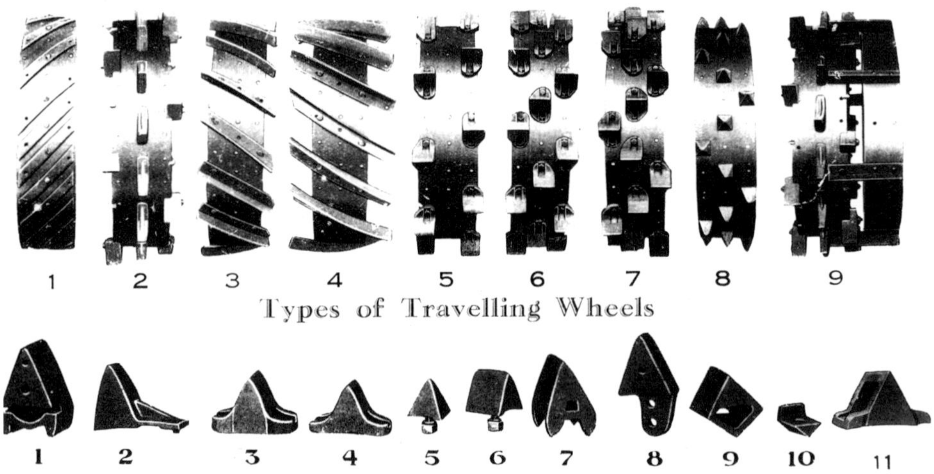

Ronaldson Bros and Tippet advertised 20 wheel and lug options for their tractors around 1930.

could be bothered with having to change cleats between jobs.

Not shown by Ronaldson-Tippett in their broadsheet was the skeleton or 'tiptoe' wheel. This was an option on some early US-built tractors. Surprisingly enough, this type of steel wheel is one of the few that is still used today. In certain conditions it succeeds where even rubber tyres fail because of its mud-shedding ability. The narrow tread is especially useful for light-tractor cultivation down the rows between vegetable crop rows.

The Allis-Chalmers Model U tractor, originally produced as the United in 1928, was made famous when the company chose this model to launch their new air tyres in 1931. The tyres were developed by Firestone from low-pressure aircraft wheels.

Rubber Tyres

Rubber tyres were rapidly adopted soon after they first appeared commercially in the US in the early 1930s. Actually the first rubber-tyred tractor in the United States was probably Thomson's Rubber Tired Steamer, which was a British steam traction engine with solid rubber-cleated wheels. It was entered in the California State and San Joaquin Valley Fairs in 1871. It could speed along at 10 mph (Gray, 1975).

The 'peculiar use of rubber' according to Gray (1975) that hastened the adoption of the pneumatic tractor tyre was the adoption of discarded truck tire casings by Florida citrus orchardists to prevent steel wheel lugs from damaging the roots of their orange trees. Such was the demand for discarded truck tyres that were being shipped to Florida in 1928 that it attracted the attention of tyre manufacturers. B.F. Goodrich's first commercial response was to release a zero pressure rubber arch tyre for Florida orchardists in 1931.

By 1932 several tyre companies were experimenting with pneumatics for tractor applications. The Firestone Company fitted an Allis-Chalmers tractor with low-pressure aeroplane tyres, then in the latter part of 1931 Allis-Chalmers launched their first Model U with pneumatic tyres as standard. This was considered a breakthrough for the time. Both Firestone and Goodyear released a range of pneumatic tractor tyres in the US in 1932.

Steel wheels had many drawbacks by comparison with pneumatic tyres:

- Limited forward speed. Since they were first intended merely to replace the horse, early tractor forward speeds were similar: 2 to 2½ mph. Uppermost tolerable speed for driving with steel-lugged wheels was around 4 mph.
- They caused discomfort for the operator and damage to roads.
- Steel lugs eventually wore out, and performance declined accordingly.
- Steel wheels tended to plug with mud.
- No single lug configuration suited all soil conditions.

Most important of all was performance: tractors with pneumatic rubber tyres could outperform steel wheels for fuel economy, tractive efficiency, speed, and overall field performance in most side-by-side comparisons under most field conditions.

Overcoming tradition presented challenges for tractor salesmen, as in this incident, related about a McDonald tractor field man, in the book on A.H. McDonald:

'Some farmers had second thoughts even after ordering tractors on pneumatics. One of the earliest instances involved Roll Brothers of Kenmare in the Victorian Mallee. Ron Reid was instructed to deliver the tractor, start and demonstrate and of course to collect payment. All went well, the clients seemed happy with the tractor performance until the day Ron packed up to return home and asked for the cheque. "Sorry," they said, "we've changed our minds. What will happen to the tyre if we run over a mallee root?" Ron was astounded. No way could he go back without the cheque. He decided there was only one thing to do. He lined up a row of mallee roots and ran the tractor over them and prayed that the tyres would not puncture. No one had done this before. The gamble paid off; there was no sign of damage to the tyres. The clients were convinced and settled up. They were enthusiastic McDonald supporters for many years afterwards.

'A week later A.H. McDonald heard about this and called Ron into the office to explain, which he did, emphasising that the order may have been lost if he had not taken the punt. A.H. severely reprimanded him for doing such a foolish thing. "What would you have done if you had ruined the tyres, and who would pay for the damage?"

'Ron pointed out that sooner or later someone was going to run over mallee roots but until now we didn't know what would happen. "Now we do know we can have a lot more confidence in selling tractors with pneumatics," he told his boss' (McDonald, 1988).

Among the local manufacturers, Ronaldson-Tippett and McDonald were quick to offer the pneumatic-tyred wheel option in the mid-1930s. Dunlop Australia had pneumatic tyres available for tractors by 1934. The earliest McDonald drive wheels were fabricated simply by welding the new rims for pneumatics onto modified steel wheel hubs and spokes.

Actually, the Australian manufacturers were obliged to offer rubber tyres since imported tractors so equipped would have left them with no market at all. Nevertheless steel wheels were still available as an

Fortescue's Universal Octopus Wheel for tractors, as advertised in Power Farming, *June 1930.*

The driver needed to keep his elbows in if he was using the Lewin Tractor Grip. A set cost £25 10s. in 1932.

'The McGregors are coming.' McGregor Grips advertised in Power Farming, *25 August 1932.*

option even into the 1950s. Today there are places and specific soil conditions where 'tiptoe' steel wheels have a unique advantage, for example, in tractor cultivation between the narrow rows of vegetable crops. In ricefields, steel strakes and cage wheels still have a useful place.

Power Farming*'s sometimes acerbic cartoonist made the point for rubber tyres in this series of illustrations in the issue of 25 March 1936. The accompanying article made a powerful case for Goodyear Farm Implement tyres.*

The Last Word for Steel Wheels

In September 1949 H.S. Brennan wrote in a letter to *Power Farming*: 'I found on steel wheels that a 10-disc Sundercut was a good second gear load (sometimes first) and the slip especially on ploughed ground, quite noticeable. Pulling the same machine in the same paddock and using rubber tyres, I find that up to fourth gear is possible on level going — about 6 mph with the large tyres, as against 2¾ mph on steel wheels. The Sundercut, however, will not stay in the ground at 6 mph ... One advantage (apart from the initial purchase price) which steel wheels have, is the ease with which they can be released from a bog, as compared with rubber, because a pole can be placed under the spikes to allow the tractor to lift itself out.'

Nowadays, Australian farmers are to a large extent obliged to use tyre patterns and rim sizes that have become standardised overseas, although the major makers do their own tyre moulding in Australia for smaller sized tractor tyres. The trend towards the traction-improving radial tyres has been entirely dependent upon imported radials.

Crawler Tracks

If only a tractor could roll along on steel rails! This was the goal of many an inventor since the first such patent for a tracked vehicle was lodged in England in 1770. R.L. Edgeworth's 'portable railway or artificial road' moved along with any carriage to which it was applied. The mechanism continuously picked up and laid down wooden rails under the wheels. The aim was to distribute the vehicle weight over a greater area than was possible with wheels and thereby allow the vehicle to traverse ground that would otherwise have been untrafficable.

In the early days of self-propulsion, few devices had the commercial success enjoyed by James Boydell's wheels. Boydell first patented the concept in 1846 then improved upon it in 1854. A contemporary account stated that the Boydell system: 'up to the present time [1860] can be regularly and profitably worked over the same ground without injuring it, and is the only engine on this principle which has ever been regularly worked' (*The Engineer*, 10 August 1917).

Crawler track was conceived in Britain and developed there and then in parallel in the US and initially known as 'portable railways'. Both types of traction systems — Pedrails and tracks — were at first cumbersome, wear-prone, and expensive.

These systems were considered in Australia but were too expensive compared with steel wheels. Where traction was deficient, steel wheels were fitted with traction-

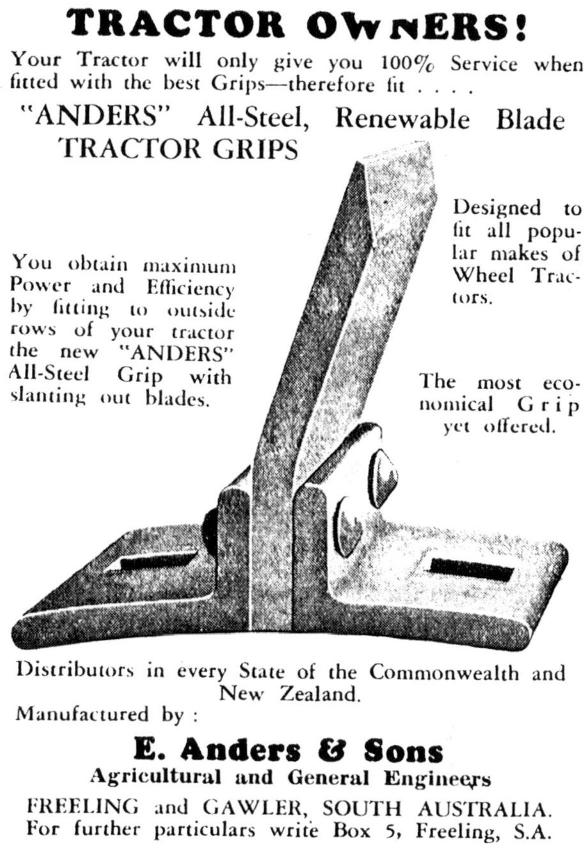

Many small (and some not so small) companies were in on the act making lugs, cleats or grips for steel wheels in the period between 1920 and World War II. Four advertisements selected from Power Farming *over that period.*

Farm Mechanics, Nov. 1938

improving lugs. Where flotation was the problem, the wheels were made wider.

The expense and complexity of crawler tracks generally caused Australian tractor makers to shy away from this otherwise very efficient means of transmitting power to the ground. One attempt by McDonald & Co. was their Steel Horse — a single-tracked machine — but it only lasted for one year, in 1917. Cliff Howard had tracks on one of his sugarcane harvester prototypes in 1931. More recently, Geoff Williames of Warragul developed and marketed a sophisticated flexible track for his self-propelled custom-built tea harvesting machines.

There was also the Altrack flexible tracked vehicle developed in Western Australia in 1988. However, the Altrack never got past the prototype stage. James Nagorcka built rubber-tracked Waltanna tractor models and sold track componentry to Deere for their ricefield combines for a while.

The tractive efficiency of a crawler tractor is between 65 and 85%, depending an soil conditions. Compare that with a two-wheel-drive rubber-tyred tractor with tractive efficiency between 45 and 80%, and for four-wheel-drive rubber tyres 55–82% (G. A. Brown, personal communication, 1987).

Rubber-tracked ATV concept machine from Altrack Limited, Perth, WA.

7 THE POWER TAKE-OFF (PTO)

Actually, most things that men do on a farm mechanically and by hand are easier than washing clothes if you haven't got modern conveniences, and less boring than cleaning the house.

— *Myra Robilliard, in* Boots and All, *1984*

An International Harvester Company artist envisaged this horsepower pto in 1928, although he hadn't quite figured out how to couple it up to the driveshaft.

The 1920 Farmall tractor was equipped with one of the earliest production model ptos.

Mention the acronym 'pto' (power take-off) to any old-timer who was involved with IH equipment in the US and they would probably respond with 'Bert Benjamin'. Bert Benjamin was 'the father of the Farmall tractor' and the 1920 version was equipped with a power take-off.

This was the first tractor to go into serial production equipped with a pto. But it was by no means the first tractor ever to use a pto. The French would probably like to claim that with their Gougis three-wheeler, which in 1907 incorporated a pto drive for implements towed by the tractor. The Gougis tractor was demonstrated working a grain binder, but it was not put into serial production (Cawood, 1984).

Development of the pto. Top: horse power or horse works, with tumbling rods, 1852. Middle: Bell-type reaper made by Aveling & Porter of England with live drive pto in 1876. Lower: G.S. Berry's steam-powered combine outfit used a live steam pto in California between 1883 and 1888.

However, the pto goes further back than that. A Bell-type reaper was put on show at the 1878 Universal Exposition in Paris. The artist's drawing of that outfit clearly shows a front-mounted reaper ahead of a steam traction engine, with the Bell-type reaper live-powered by a system of chain drives from the steam engine.

As if that isn't precedent enough, on the other side of the Atlantic, combine harvester builders piped steam from the boiler of the primary traction engine by flexible hose to an auxiliary steam engine mounted on the combine. This was a 'live steam pto'. George Stockton Berry did it first in California. He was also the first in the world to build a self-propelled combine harvester — in 1883. Berry was also shrewd enough to utilise the straw as fuel for the steam boiler. The threshed straw of this self-fuelled machine was discharged from the separating section of the combine onto a chute, then delivered to a trailer platform adjacent to the bailer where it was stoked into the boiler grate for fuel. A supply of straw would be stacked on the trailer rack at night ready for the 30-minute task of getting up steam the following morning. As the rack filled up during the day's harvest, any excess straw would be dumped onto the ground. These straw dumps were used later as fuel for the same traction engine when it was disconnected from the harvester and used for tillage. Wood and coal were scarce in the Benicia district of California.

Ronaldson-Tippett's tractor power take-off attachment.

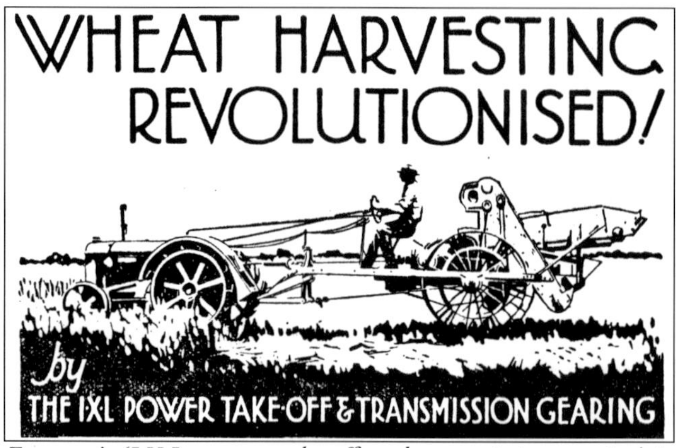

Fortescue's 'I.X.L. power take-off and transmission gearing' in 1928. (Power Farming, *25 August 1928)*

Australian Power Take-offs

The legendary Cliff Howard was probably first to commercially exploit a power take-off on an Australian machine. Howard's earliest Rotavators used a chain-drive pto and he took out patents on his 'drive means' for rotary tillers. His company also offered a chain-drive pto kit for use on tractors hauling header-harvesters, thereby eliminating the less efficient ground-wheel-drive system.

Geo. E. Fortescue and Sons Ltd, of Sydney, featured vigorous advertising in the rural press for their products in the 1920s. Among their product lines was a live pto conversion kit for tractors. It was available first for the Fordson around 1926. By 1928 they claimed to have designed power take-offs for 'all popular makes of tractors', along with 'transmission gears to convert almost any header to a power-driven machine, forming a flexible unit capable of turning in less than its own

Howard's chain-drive power take-off system.

length, giving at least 40% saving in fuel and time. The saving of grain in down crops is astounding: it has been proved to save the cost of the outfit in a few days work.'

By the 1930s tractors were being manufactured with the pto either as standard or as an optional extra, so the need for a retrofit package faded away.

Nowadays all tractors are fitted with power take-off as standard equipment, and there is more than one pto on many tractors.

Coopers offered this hand-powered pto for shearing in 1914.

Advertising from International Harvester Company of Australia, which promoted their ground-speed power take-off. (Power Farming, *March 1925)*

Howard Auto Cultivator's Fordson power take-off attachment, 1927. One wheel removed for clarity.

8 TRACTOR TESTING IN AUSTRALIA

A bill to provide official tests for gas, gasoline, kerosene, distillate, or other liquid fuel traction engines in the State of Nebraska and to compel the maintenance of adequate service stations for same was introduced by Wilmot F. Crozier before the State legislature, after Crozier had reasoned, 'If one manufacturer can build a dependable tractor, why can not all tractors be dependable?'

This bill was passed in 1929 and became known as the Nebraska Tractor Test Law

— *Cited in Gray, 1975*

Early Tractor Field Trials

As long as farm machinery has been around, there has been public interest in machinery field days — and in seeing how new equipment performs in comparative tests. The early development of tractors was accompanied by a number of public trials to demonstrate their capabilities.

Before the Nebraska test scheme (mentioned in the opening quote) was established there had been other less formal tractor trials by independent bodies. The most notable were the Winnipeg Motor Competitions, from 1908 to 1913 in Canada; and the Royal Agricultural Society of England trials, which started in 1910.

In Australia, the Royal Agricultural Society of Victoria arranged for a series of 'motor tractor' trials at the State Research Farm, Werribee, in connection with their Annual Show, held in September 1918.

The tests covered both ploughing performance and engine performance. Engine performance was reported on by Professors Payne and Kernot and E.J.C. Rennie from the University of Melbourne. The tests were made using a belt-driven dynamometer at the Engineering School. Power and fuel consumption were measured. Ploughing performance was evaluated by Mr Ross of the State Research Farm, Werribee. Field work rate and fuel consumption were measured, as was drawbar pull, using an autographic dynamometer. Average ploughing horsepower was calculated for each of the tractors involved in those first trials, which were:

- Bates' Steel Mule, distributed by Clutterbuck Bros,
- Waterloo Boy, distributed by Mitchell and Co.,
- Sunshine Models A and O made by H.V. McKay,
- Imperial FAA and Imperial Lightweight, made by A. H. McDonald and Co.,
- Mogul, from International Harvester Co., and
- Crude Oil Tractor, made by Jelbart Pty Ltd.

The results were reported in the December 1918 edition of the *Victorian Journal of Agriculture.* The trials did not become a regular feature of the Melbourne Show, though simple tests of pulling capacity of horses, much like later tractor pulling competitions, were a feature of the show until well into the 1930s.

Origin of the Australian Tractor Testing (ATT) Scheme

The Australian Tractor Testing Scheme officially began in 1946 and ceased on 30 June 1976. The scheme was established by the Commonwealth government, initially to meet its own need for information on aspects of tractor performance that were used in assessing federal duties, taxes and bounties. A listing of all the tractors tested under the ATT scheme follows.

The annual Winnipeg trials continued for five years, 1908–13, and attracted international attention. A brake dynamometer tested steam traction engines and I-C tractors for belt power.

Australian Tractor Test Listing

TEST NO.	MAKE AND MODEL OF TRACTOR	BELT OR PTO POWER hp	DATE Month/Year
1	Chamberlain 40K	42.7	3/46
2	Fordson Major - Standard	32.5	7/46
3	Fordson Roadless	32.5	7/46
4	Fordson Major with Perkins P6 Diesel	41.0	3/49
5	Ferguson TEA20	18.8	5/49
6	Fowler Marshall Mark VF	36.0	6/49
7	Newman WD2	7.5	7/49
8	David Brown Cropmaster VAK/IC-65	19.6	9/49
9	Minneapolis - Moline Twin City type UDS	42.5	9/49
10	Anzani Iron Horse	4.3	11/49
11	Deutz Diesel F3M317	42.6	11/49
12	K. L. Bulldog	44.2	3/50
13	Hanomag Diesel R40	38.6	4/50
14	Hanomag Diesel Tracked KV50	31.1	5/50
15	Lanz Bulldog	43.0	6/50
16	Nuffield M4	31.0	6/50
17	Lanz Bulldog P	46.9	7/50
18	Renault 3042	23.8	8/50
19	Fordson Major County Crawler - Diesel	43.0	10/50
20	Fordson Major County Crawler - Kerosene	30.6	11/50
21	Renault R.3042	26.4	6/51
22	Sift Diesel TD4	38.0	10/51
23	Chamberlain 40KA Kerosene	41.5	10/51
24	Fordson Major - Diesel Model	39.4	7/54
25	Fordson Major - Kerosene Model	32.3	7/54
26	McCormick International Super AWD-6 (Diesel)	47.6	6/55
27	McCormick International Super AW-6 (Kerosene)	41.3	7/55
28	Zetor Super Diesel	41.2	8/57
29	Zetor 25A Diesel	24.5	9/57
30	International AW-7 Diesel	48.8	11/57
31	David Brown 900D	34.6	1/59
32	Fordson Dexta	30.0	7/59
33	Fordson Farm Major	39.8	9/59
34	Fordson Power Major	34.7	5/60
35	Zetor 50 Super	51.0	9/60
36	International A554D	54.0	12/61
37	Chamberlain Canelander	43.3	5/63
38	International A414	36.1	3/65
39	Ford Dexta 2000	31.0	5/65
40	Fiat 415	37.6	9/65
41	Massey Ferguson 135	31.8	11/65
42	John Deere 1010RS	32.0	2/66
43	David Brown 880A	37.4	6/66
44	Ford Super Dexta 3000	34.9	8/66
45	Nuffield 10/42	30.2	10/66
46	International 564	53.1	5/67
47	International 434	33.5	6/67
48	David Brown 990	44.9	8/67
49	McCormick International 624	53.0	9/67
50	Ford Major 4000	44.0	10/67
51	Ford Super Major 5000	59.0	11/67
52	Chamberlain Champion 306	57.6	2/68
53	Chamberlain Countryman 354	75.0	3/68
54	Fiat 615	54.0	6/68
55	Nuffield 4/65	50.3	9/68
56	Zetor 5511	48.0	12/68
57	David Brown 1200	54.6	4/69
58	Massey Ferguson 165	53.2	9/69
59	Massey Ferguson 178	61.2	9/69
60	Fiat 650	58.0	11/69
61	Same Centauro 60DT 4WD	45.4	12/69
62	Same Centauro 70DT 4WD	53.3	12/69
63	John Deere 2120	55.2	5/70
64	John Deere 2020	46.3	9/69
65	Fiat 550	42.7	12/70
66	Chamberlain C670D	58.7	2/71
67	Ford 4000	47.3	3/71
68	Ford 5000-6Y	62.5	7/71
69	Ford 8000	94.7	8/71
70	International 686	65.5	9/71
71	International 756	72.2	9/71
72	Chamberlain C6100	79.6	11/71
73	John Deere 1120	41.6	8/72
74	Fiat 750	51.8	9/72
75	International 574	50.0	11/72
76	Chamberlain 236	52.7	1/73
77	Ford 7000	76.8	3/73
78	Leyland 253	28.0	9/73
79	International 444	27.0	6/73
80	International 696	49.0	6/73
81	Massey Ferguson 148	28.9	8/73
82	Leyland 270	41.5	4/74
83	Leyland 255	33.6	6/74
84	Fiat 1000	62.5	2/75
85	International 766	59.8	5/75
86	International 866	70.7	5/75

Rapid Expansion in Tractor Population

A second important reason for setting up the Australian testing scheme was that the rapid expansion in the tractor population after World War II had created interest by some firms in designing and manufacturing tractors in Australia, and in having their performance claims supported by an independent authority in a similar fashion to that offered by the Nebraska Tractor Test system. The Nebraska tests were established in 1919 when the legislature in that state passed a law requiring anyone intending to market a tractor there to have a permit issued by the State Railway Commission. The permit could only be issued after a stock model had been tested at the University of Nebraska and the performance of the tractor compared with the claims made by the manufacturer.

The experience of Wilmot F. Crozier, a pioneer US tractor farmer who introduced the Nebraska bill, is relevant and is cited by Gray (1975) this way: 'The first tractor he bought [in 1916] was called a "Ford" but had no connection with Henry Ford. It gave him so much trouble that he demanded a replacement. The replacement in 1917 was no better than the original. In 1917 he purchased a secondhand Bull tractor. This was unsatisfactory. In 1918 he purchased a three-plow Rumely Oil-Pull. He pulled five bottoms satisfactorily with this tractor which gave him absolutely no trouble.'

The Nebraska law was introduced to give farmers some protection from the exaggerated performance claims that were being made by some manufacturers. Tractors began to be generally accepted in the US after World War I and many new companies, with little experience, were seeking a share of the burgeoning market. The requirement to have the tractor tested before it could be marketed in Nebraska had the effect of rapidly eliminating less responsible manufacturers and gave buyers a guide to the

Bates' Steel Mule, an American tractor distributed by Clutterbuck Bros of Melbourne, on test at Werribee in 1918. (Photo supplied by Martin Hallett, Museum of Victoria)

The International Mogul tractor. An IH Mogul was tested at Werribee in 1918. It is shown parked behind the Engineering School of the University of Melbourne. (Photo courtesy Department of Agriculture: Victoria)

Diagram of 8-16 HP IH Mogul tractor used in the company's advertising.

performance that might be expected. The program also hastened improvements in tractor design.

Most tractors brought into Australia before 1940 came from North America, so Nebraska performance information was usually available. However, after the war, tractors were increasingly imported from Europe (where formal published tests had not commenced) or had a significant Australian content, so there was less independent test data on those tractors.

The Early Years of the Formal Testing Scheme

The testing scheme was initially run by the Aeronautical Research Laboratories (ARL) of the Commonwealth's Council of Scientific and Industrial Research (CSIR). The testing laboratories were located at Fishermens Bend near Melbourne. Engine performance was assessed using a belt-driven dynamometer, and drawbar performance of rubber-tyred tractors was measured on a bitumen or concrete test track. Endurance testing was conducted around the Albert Park Lake. For crawler tractors a test strip was found on the State Research Farm, Werribee.

The first official test took place in March 1946. The tests were conducted by R.A. Wright of ARL and G.H. Vasey of the Agricultural Engineering section of the School of Agriculture at the University of Melbourne. Inspection of the tractor after test was made on behalf of the University of Melbourne by Mr M. Shaw and Mr E.J. Rennie, the same Rennie who had been involved in the 1918 trials.

That first test was on a prototype Chamberlain 40K tractor made by the Australian Ball Bearing Co. Pty of Salmon Street, Port Melbourne. The kerosene-fuelled tractor was rated at 30 drawbar horsepower and 40 belt horsepower. In the test the prototype produced 42.7 HP at the belt pulley and 41.0 HP at the drawbar. The engine had two horizontally opposed cylinders and ran at a rated speed of 1200 rpm. Fuel economy on the belt test was 0.69 lbs/HP.hr.

Drawbar load for this test was measured using a Watson recording drawbar dynamometer. The load was provided by towing three large motor trucks. Each truck had a five-speed gearbox and by towing the trucks in various combinations of gears the required degree of load was obtained. Wheel slip was found by measuring the distance travelled in 20 revolutions of the driving wheel.

The Chamberlain 40K had nine forward gears. At that time most tractors had a gear range of only three or four. This led to some testing problems, as shown in the following extract from the test report: 'It was possible to test the tractor in 3rd, 4th and 5th gears only, as the lower gears would have imposed loads beyond the capacity of the dynamometer. In higher gears wheelslip could not be measured, as the observer (running alongside and counting wheel revolutions) could not keep up with the tractor at the faster road speeds, and no automatic counter to obtain wheel r.p.m. was available.'

The test procedure followed was identical with that used at Nebraska. The test was reported as

The earliest official Australian tractor test, later designated as ATT Number 1, was that performed on the Chamberlain 40K in 1946.

The Test Station's first auto-dynamometer.

Chamberlain 40K on belt-pulley test driving the water absorption-type dynamometer. (Photos supplied by Martin Hallett, Museum of Victoria)

Engine Technical Memo No. 17 of the Division of Aeronautics.

Test numbers 2 and 3, on a Fordson Major and Fordson Roadless respectively, were carried out using similar test equipment in 1946. By test number 4, reported in March 1949 (on a Fordson Major fitted with Perkins P6 diesel), a new hydraulic dynamometer with gauge readout had been designed and built by ARL and this was mounted on the tractor load vehicle. Wheel or track slip was measured by using an electric counter to record the number of driving-wheel revolutions over 500 feet to an accuracy of one-14th of a revolution, so overcoming the problem of forward speed in the higher gears.

Test number 5, on the Ferguson TEA 20 in May 1949, brought the first divergence from the Nebraska procedure. Belt power tests at 'maximum load', at 'rated load' and at varying loads remained the same. 'Rated power' was defined as 85% of maximum belt power. Drawbar power was measured at maximum load in rated gear, the gear nearest to 3.5 mph, and at maximum operating load in each of the gears. These tests were made in accordance with the Nebraska procedure. However, in the endurance test, the Nebraska procedure called for over 10 hours' testing at rated load in the rated gear. From test number 5 on, the Australian testing body chose to divide the 10 hours into four periods of two and a half hours each at a different load so that fuel consumption figures under the four load conditions were obtained. The loads ranged from 90% of maximum drawbar down to 37% of maximum drawbar power.

The ubiquitous 'Fergy' tractor (TEA 20) was tested at Werribee in May 1949 (ATT Number 5). The Ferguson TO-30 model of 1961 vintage is shown.

By 1950, R.A. Wright was able to report that two dynamometer cars of different sizes were available (Engine Tech. Note 153). Loading was provided by an oil pump driven through a two-speed gearbox by the test truck wheels when the truck was towed by the tractor. A valve in the discharge line of the pump allowed control of the amount of power absorbed by the pump. Extra loading could be obtained by placing the truck in gear and by towing in gear additional trucks or tractors behind the loading car. If a very light load was required, the truck engine was started and used to overcome part of the load vehicle's rolling resistance.

In 1949 ARL was transferred to the Department of Supply, and in 1951, after 23 tractors had been tested, departmental changes led to a suspension of the scheme. The last tractor tested by ARL was the Chamberlain 40 KA model, in October 1951.

The Tractor Testing Agreement

The University of Melbourne had been involved right from the inception of the testing work and so it was logical that the university should be approached to take over the testing role when the ARL could not continue. The continuation of the testing scheme was organised through the Australian Agricultural Council, so involving the Commonwealth and all the state governments.

A Tractor Testing Agreement was entered into between the several governments and the University of Melbourne. In the early stages the Bureau of Sugar Experiment Stations acted on behalf of the State of Queensland. The agreement called for professional services to be provided by the university and for the governments to meet the costs, initially half from the Commonwealth and half from the states, in proportion to their tractor populations. Over the years this sharing of costs varied, both in those participating and in the proportion of cost contributed by the various parties.

Under the agreement the university would carry out tests and also research work on tractors and related matters. The scheme was to continue as a voluntary one — there was no compulsion to submit tractors for testing— and was expected to recoup some of its expenses by charging test fees (previously there had been no charge for the tests). To manage the scheme a Tractor Testing Committee was established, made up of an officer of the Commonwealth Department of Primary Industry as chairman, an officer of the Victorian Department of Agriculture representing the interests of the states, and an officer from the University of Melbourne.

Testing restarted in 1954 using temporary premises at Fishermens Bend. Permanent buildings were built on the State Research Farm, Werribee, and officially opened in 1957 by the then Minister of Primary Industry, Mr William McMahon. G.H. (Gilbert) Vasey, by then a Senior Lecturer in Agricultural Engineering in the Faculty of Engineering, was the first officer-in-charge of the Tractor Testing Station. A tractor testing officer, Ian Naylor, was appointed in 1954 and was succeeded by W.F. (Bill) Baillie. In 1957, a technical officer, Norman Walkadine, was appointed and continued until 1967, when H.W. (Harry) Duddy became the technical officer.

The basic role of the Testing Station was to carry out formal tests on tractors to provide information for purchasers. The testing protocol was continually reviewed and improved. From the original performance-only tests, the scheme gradually expanded to cover physical characteristics — including centre of gravity and turning circles — and to report on conformity with standards for three-point linkage, drawbar, power take-off and so on. Some assessment was made of comfort and accessibility. The performance measurements were expanded to include the full range of engine performance. The full test included a comprehensive set of graphs, while a simpler, abridged version, known as the Farmer's Edition, summarised the results in tables which were usually picked up by the rural press.

In the early 1960s a dispute arose with manufacturers over details of the test procedure and most particularly over the issue of the method of selection of the tractor to be tested. The Australian Tractor Testing Committee insisted on random selection, whereas at Nebraska and at the UK testing authority, the National Institute of Agricultural Engineering, the tractor to be tested was chosen by the manufacturer. The issues were fully canvassed in papers by Baillie and Vasey (1968), Manby (1960) and Donaldson (1970). Determined efforts were made by certain manufacturers to have the scheme closed down.

Eventually, when the government had made it clear that it would if necessary purchase tractors for test, the manufacturers agreed to cooperate with the scheme, and to random selection of a sample for test. The arguments at least led to a clarification of the issues involved, and the continuation of the scheme was clearly understood as a consumer testing service. At the same time, the test fee of about $500 to $600, which had been paid by the manufacturer or importer, was discontinued.

The test program, with the active cooperation of the trade, set out to test the available tractors in a particular power range. As a result of this, three summary reports were published, setting out in comparative forms the results of tests on tractors in the 34–45 HP class, the 45–60 HP class, and the 60–70 HP class. These comparative reports had wide acceptance and were reprinted overseas. Individual test reports were distributed on a mailing list and through state departments of agriculture. They were reprinted in the journals of agriculture of some states. 2000 copies of the full report were printed along with 5000 copies of the abridged version.

The renewed vigour of the testing program led to a review of the Tractor Testing Scheme. As indicated above, the contributors' enthusiasm had varied over the years, but in 1969 a new agreement was entered into for a five-year period, guaranteeing a set income over that period and providing for an extension of the buildings. The new agreement allowed the recruitment of extra staff. In 1969 W.T. (Bill) Brown joined the station as a testing officer and in 1970 B.J. Benham joined as a second technical officer. Gilbert Vasey retired from the position of officer-in-charge in June 1970 and was succeeded by W.F. Baillie.

Other Activities for the Test Station

The charter of the Tractor Testing Station allowed research into aspects of tractor performance and related issues. In 1947 a technical memo had noted the need for improved spark arresters on tractors. In 1957 a test procedure was drawn up on the basis of the standard developed by the American Society of Automotive Engineers (SAE). The Testing Station began testing to this procedure in 1957. Eventually the procedure was examined by the Standards Association of Australia, and after revision was adopted as Australian Standard 1019:1970 'Spark Emission Control Devices for Internal Combustion Engines'. The station completed 78 spark arrester tests, though only those that were successful were published. The experiences derived from the testing program were reported in a paper by Vasey and Baillie in 1961.

Beside fire risk, the committee had asked for a watching brief to be kept on tractor safety. A collection of newspaper reports of tractor accidents was begun in 1961. These reports were then analysed and revealed the high accident rate for farm tractors and the dominance of overturning as the major cause of tractor fatalities. Work in Sweden had led to the development of a test for frames or cabins intended to provide protection in the event of a roll-over. Knowledge of the Swedish procedure was used in 1964 to test a roll-over-protective-structure (ROPS) known as the Alexander Bow. Later, following the adoption of the British standard by the Australian Standards Association, the Testing Station began testing ROPS as a regular service. The new agreement of 1969 provided funds for the building of an extension to house the ROPS test rig. This required a high-bay building with an overhead crane and weighbridge, both of which made formal tractor testing much easier.

The test for acceptance of a protective frame or cabin required a prototype of the design, mounted on a tractor of the type for which it was intended, to be subjected to a series of impact and crushing tests of a strength related to the weight of the tractor. The cabin or frame is struck from the rear with a two-ton pendulum weight. This is followed by a vertical crushing test to the rear with a load of twice the weight of the tractor. Further blows to the front and side are delivered, followed by a final crushing test on the front of the frame.

The first ROPS tests were carried out on a Gason frame fitted to a Ford tractor in 1971. By 1976 the station had completed 103 ROPS tests.

Beside carrying out the testing of ROPS, staff actively participated in the promotion of the fitting of ROPS to tractors. Nowadays they are a legal requirement in all states for new tractors. The effect of legislation and propaganda is illustrated by the dramatic reduction in tractor fatalities in Victoria — from an average of about 20 down to about five a year.

A further aspect of occupational health related to tractors is that of excessive noise and hearing conservation. In 1959 noise levels around tractors were investigated at the Tractor Testing Station (Hutchings and Vasey, 1964) and, as a result, the measurement of the noise level at the operator's ear was included in Tractor Test Reports from 1962. The Australian Tractor Testing Station was the first in the world to include noise measurements in its test reports.

The station undertook many other tests and investigations. A number of prototype tractors were tested before they were marketed. Tests were made on the tractive performance of tyres, on the power of chain saws, the draft requirements of asphalt-laying machines, on engines and on radiators.

Ford 5000 tractor on test at Werribee in July 1971. The dynamometer vehicle had a capacity of 8000 pounds pull and around 210 HP, limited by the heat-dissipating ability of the dynamometer. (Photo courtesy Victorian Department of Agriculture)

An unusual test was made in 1949 by ARL, with assistance from the University of Melbourne. The test measured the traction of an eight-horse team. The names of the horses were Margot, Florrie, Dutchy, Dollar, Lubra, Searchlight, Barmaid and Flash Jess. The report concluded: 'a team of eight medium weight draught horses may be expected to exert a drawbar pull of sixteen hundred pounds at 3 mph (15 HP) throughout a working day. A pull of 2800 lb at 4 mph (34 HP) can be exerted for a short period.'

Closure of the Station

In 1974 the five-year agreement came to an end. Over that period there had been a rapid rise in inflation and tensions had developed between the Commonwealth and the States. A review was made of the Tractor Testing Scheme. Some states criticised the amount of time spent on promotion as being excessive, others said it was too little. Reports were said to be too late in appearing, and too complex for farmers. Several states felt that their financial contributions could be better spent elsewhere (the contributions from those critics were just $1500 each). Having seven Treasuries assessing the merit of the scheme did not help its prospects.

In 1975 the decision was taken to discontinue the scheme and it ceased operation on 30 June 1976. The buildings were sold to the Victorian Department of Agriculture and the three remaining staff transferred to the Victorian Public Service. This action laid the foundation of the Agricultural Engineering Centre in Victoria. The equipment was given to the University of Melbourne, which in turn lent it to the Victorian Department of Agriculture to allow some of the testing activities, mainly those associated with safety, to continue.

Perhaps the success of the scheme could be measured in the results of a survey of farmers which showed that 80% knew where the station was located while 14% said they had used test reports in selecting a tractor.

The Australian Tractor Testing Station had a big influence in the development of understanding of tractor performance, and in the promotion of safe tractor operation. Nowadays, Australian farmers are entirely dependent on overseas test information on tractors (and even that is drying up since Nebraska phased down) or on trade literature — which puts us all back to about where we were in 1919 as far as testing is concerned.

[I am indebted to Bill Brown of Werribee for the work he put into this chapter and for supplying illustrative material.]

Mallee Mike breaks in a 'crude' one. (Power Farming, *September 1927)*

Tests Completed by the Australian Tractor Testing Station, 1954–76*

Spark arrester tests	53
Roll-over protective structures	103
Noise tests	7
Engine tests	20
Other private tests	62

* *Note:* There were 23 tests conducted on tractors, three of them of Australian origin, prior to the formal establishment of the station in 1954.

INDIVIDUAL TRACTOR MANUFACTURERS

9 A.H. McDONALD & CO. — AUSTRALIA'S FIRST TRACTOR MANUFACTURER

Whatever the client wants, we can make it.
— *McDonald company philosophy*

Alfred Henry McDonald was a baker's son. His family lived behind their bakery in Glenferrie Road, Hawthorn, Victoria, and Alf helped in the shop when he left school at age 14, but his heart was not in it. He wanted to be an engineer. Despite some resistance from his father, he became indentured as an apprentice to a Frenchman, Henri Galopin, 'Scientific instrument maker to the Observatory', in Chancery Lane, Melbourne. After successfully completing his apprenticeship, he next worked briefly for an electrical engineer, and after hours did jobbing work in a shed behind the bakery. Once, at a dental appointment, he enquired why the dentist didn't have an electric motor to drive his treadle-powered drill. The dentist replied that if a suitable motor was available for his 'dental engine', he would buy one. Alf went home and soon had one built. The dentist bought it on the spot. This was the first of many McDonald electric motors supplied to dentists.

In August 1903, at the age of 20, Alf registered the business name A.H. McDonald & Co. He set up a workshop in Flinders Street, Melbourne, with his brother, Ernest John McDonald, to make electrical gear, appliances — and their first engine, the 5 HP No. IA. Setting up the workshop included assembling their own engine-driven generator to provide them with power for the works. In 1904 they moved the enterprise into a corrugated-iron shed in Burwood Road, Hawthorn, where their original No. 1A engine-powered generator set was the sole source of power. Alf sensed that a new age of power was coming to Australia and he became consumed with the idea of building engines. As his interest in engines grew so his interest in electrical products diminished. In 1905 he commenced a production run of 30 A-type engines of 4 HP per cylinder, offered in two-, three- and four-cylinder models. Then in 1907 came the larger D-type engine, of 6.25-inch bore and 10 HP per cylinder. A twin-cylinder D-type engine was the one selected for Australia's first tractor, which was built in 1908.

A.H. McDonald & Co built Australia's first locally designed tractor. Right from the start, company principal Alf McDonald, who was ever on the lookout for equipment that could be powered by his engines, perceived that the steam traction engine was not a suitable power source for Australian farmers. And he saw a future in farm tractors.

Alfred Henry McDonald (1883–1963), governing director and founder of A.H. McDonald & Co., builder of Australia's first locally manufactured tractors, which he commenced assembling in 1908.

That first McDonald Imperial oil tractor was sold to a farmer in July 1909. It was used on land near the Geringhap railway station, Batesford, Victoria. The purchaser, Mr J.H. Dardel, was impressed with 'instantaneous' starting ability of the McDonald Imperial: he said it took but 'five minutes to be got ready — there being no loss of time in getting steam' (*The Fruit World of Australia*, November 1909).

Compared with lumbering steam traction engines, this was quite a step forward, and at 4.5 tons Dardel's

The first commercial Australian-built tractor, an Imperial Oil Tractor Model EA by A.H. McDonaid & Co., as exhibited at the Royal Melbourne Show in 1909. It operated on kerosene.

tractor was a mere lightweight. Dardel's McDonald Imperial tractor could cover 13.5 acres/day pulling a four-furrow disc plough. It used 1.47 gallons of kerosene, at a cost of 1s. 7d. per acre.

The first tractor, a Model EA, with its 20 HP four-stroke engine mounted crosswise on the chassis, had magneto and coil ignition, and was fuelled by kerosene after being started on petrol. Thirteen EA tractors were built between 1908 and 1911.

The earliest cooling system employed was simple but inefficient. A plunger-type water pump circulated water through the vertical engine jacket to the top of a cooling tower where pipes drilled with small holes allowed the water to dribble down to the main water reservoir at the base of the tower. The engine's exhaust was piped to the top of the cooler where its vertical outlet created a draft of air, taken in through holes placed just above water level. On the earliest engines, these holes were at first placed too low and resulted in severe loss of water, so they were sealed off and new holes drilled higher up.

The Model D engine used in the McDonald tractors had an extended crankshaft drive to the gearbox through the clutch pulley. The idea of using the clutch pulley to transmit the power through a gear or chain drive to the transmission was continued by the McDonalds for many years, in both their tractors and road rollers. A belt pulley was essential for driving other machinery, such as chaffcutters and pumps, and for stone-crushers. For belt work the gear selector was placed in neutral and the clutch lever was used to engage or disengage the drive pulley.

The transmission provided three forward speeds and one reverse. The earliest gearboxes were well supported in a cast iron housing but it had no bottom — the gears did not run in an oil bath. There was a removable access cover on top and oil cups for lubricating the journal bearings. The driver simply lifted the access cover from time to time and used an oil can to squirt lubricating oil over the gears.

The differential was housed in a separate casing to keep dirt out and this casing provided an oil bath for lubrication. The final drive was through gears to the steel wheels.

By 1910 the company had outgrown the Hawthorn premises, so they moved again,

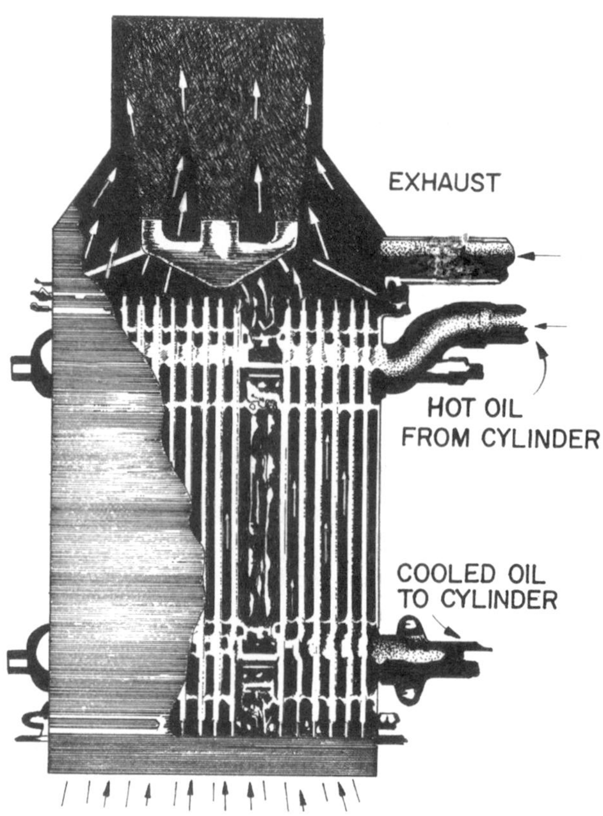

The early McDonald tractor induced-draft tower engine cooling system was a variant on the systems used by such US makers of the time as Hart-Parr and Rumely, who used oil as the cooling medium, whereas McDonald's engines were water-cooled.

this time to a factory in Stawell Street, Burnley. Here they added a foundry with unit casting capacity of up to three tons. The product line was expanded to include a more powerful engine series, the N type verticals, of 20 HP per cylinder (8-inch bore), again offered in two, three and four cylinders.

The next year was an adventurous one for the company, when the famous McDonald road rollers went into production. Also an enormous tractor was contract-built for constructing irrigation ditches. The driving wheels of this special tractor were 11 feet in diameter and the front wheels 6 feet 6 inches. The most unusual feature was the mounting of the engine (said to have been rated at 120 HP) over the front wheels, while the bucket elevator of the ditcher was mounted between the rear wheels for discharging the material dislodged by the blades. Outriggers were suspended by cables hitched to a mast protruding above the tractor canopy. And 1911 was also the year that Frank Bottrill came to see the McDonalds.

1911 was a banner year for the McDonalds, who launched their famous Imperial road roller series. They also contracted to build this large tractor for irrigation-channel construction. The four-cylinder engine design of around 100 or 120 HP was probably modelled on the McDonald suction gas engines of the time and built especially for this tractor. The doors of the factory had to be removed to get this machine out of the works.

The Bottrill Connection

Frank Bottrill, who later built Australia's (physically) biggest tractor, Big Lizzie, had an idea. He would improve the flotation and traction of heavy and cumbersome tractors by exploiting planks suspended on each wheel which the wheel would then drive over. He called them 'Dreadnaught' wheels, others called them 'Pedrails'. He sold his first set to Mount Gunson copper mines in South Australia. The company had called tenders for a special tractor with a trailer, each equipped with Bottrill wheels. The tender was awarded to Austral Otis. In August 1911 the tractor, with its 40 HP twin-cylinder McDonald engine and Pedrails, was ready for testing on swampy flats behind the Austral Otis works in South Melbourne.

After passing these trials, the tractor was dismantled and shipped in no less than 200 separate packages to Hummock Hill (now Whyalla). It was met there by Frank Bottrill and his brother Reuben, who assembled it and drove the rig to Port Augusta. The tractor worked successfully for the mines for a number of years. The McDonald company offered Bottrill's Pedrails as an option on farm and industrial tractors of the period, as well as on a special version, the Model EC tractor.

Dardel's tractor, the original, was returned to McDonalds for a complete rebuild in early 1912. On the delivery run following the completion of the rebuild, the driver reported that 'by the time we reached Power Street the spur wheel connected with the steering column came off no fewer than three times. We had not gone far before the tool box (with seat) came off and I went with it and had a narrow escape from being run over'. Dangerous work! A.H. McDonald's memo to E.J., the works manager, says this episode 'clearly indicates there is room for a great deal of improvement with respect to the supervision of the work before it is sent out'. There was tension and even acrimony at times between the hard-

E.J. McDonald, brother of A.H., beside an N-type engine.

driving Alf and the more retiring E.J. McDonald.

In 1913, an order was received from the Australian Transport Co. for two tractors to carry stores over uncharted terrain to mine sites. These 30 HP three-cylinder-engined McDonald tractors were demonstrated at Fishermens Bend, where the industrial press acknowledged that difficulties with regular steel wheels sinking into soft ground were 'surmounted by the Bottrill patent wheels'.

In November 1913, the Commonwealth Railways, which were building a line across the Nullarbor Desert, announced that: 'in order to expedite the transport of water and material along the route of the East-West railway and to reduce the cost, an oil tractor and trailer has been ordered by the Assistant Minister for Home Affairs from Messrs A.H. McDonald & Co., Burnley, Victoria and should arrive at Port Augusta shortly'. McDonalds were awarded contracts to build Pedrail-equipped tractors, one of 20 HP and one of 30 HP, and were able to successfully demonstrate the efficacy of their tractors in a series of trials at Fishermens Bend, where the sandy conditions were considered akin to those of the Nullarbor. Standard-wheeled tractors bogged down quickly in the sand and could not be moved. The Pedrail tractor clumped on over the sand and swamp. A steel cable was hitched to the bogged wheeled tractors and they were hauled out without difficulty. The Commonwealth Railway's Construction Branch officials were satisfied, and the tractors were duly shipped to Port Augusta.

Reuben Bottrill went out to commission these tractors for the trans-Nullarbor rail link. At one stage in 1914 he wrote to his brother: 'When at the 132 miles [mark] I had an inch of rain to contend with and at the time it came on there were two large claypans (like glue) and a salt lake to contend with too — directly in front of me. For the next two days I travelled only four miles, but I got past the claypans and lake and got into heavy sand with the one mile gear on all the time for about 12 miles. The last three days of return trip I did 13,17 and 15.5 miles respectively. I think this tractor has proved itself very suitable to the Com' during this trip, especially after getting past those gluepots with an inch of rain there'.

The tractors proved eminently suitable for the task of establishing camps and water bores over trackless desert, keeping ahead of the main rail works. On completion of the job, the tractors were apparently left where they had last worked, and were eventually covered by the drifting sand in distant South Australia.

McDonald tractor equipped with Bottrill wheels for the Transcontinental Railway route in 1913. Before the Model EC with Pedrails was shipped to South Australia, it was tested on sand at Fishermens Bend in 1912.

More will be heard about Frank Bottrill and his tractors in a later chapter. Suffice to say that the association with Frank Bottrill wound up with Bottrill's total commitment to his ultimate tractor, Big Lizzie — a machine built largely by the McDonald works, but fitted out by Bottrill with a Blackstone 60 HP engine.

Moffatt's Scrub-clearing Tractor

In 1914 the McDonalds were approached by a contractor, H. Moffatt of Warragul, who wanted a powerful machine to clear scrub in Gippsland. In keeping with their philosophy to build 'whatever the client wants', they produced their Model EE Special, with four-cylinder DB engine of 40 HP, mounted lengthwise in the chassis over the rear axle. This time the enormous engine cooling system was built like a water-tube boiler, with a large fan drawing air through the cooling tubes

McDonald Imperial Model E-B tractor carting wheat to railside after harvest. The owner was J.A. Troup, of Tourello, Victoria, who boasted of ploughing 500 acres at a cost averaging between 3s. 9d. and to 4s. per acre, compared with horses at a cost of 10s. per acre.

In 1914, this model EE special-purpose tractor was built for contractor H. Moffatt for land-clearing work pulling a heavy-duty scrub roller near Warragul, Victoria. It used the 40 HP Model DB engine, with characteristic forced-draft water tube cooling.

from front to rear. This fan system broke with the tower-cooling system, which McDonalds kept working at in many variants up till 1916.

The Steel Horse

In 1916 the McDonalds planned on building the Steel Horse series. These were to be lightweight tractors, with a single driving wheel at the rear and two wheels for steering spread well apart at the front. They were driven by a single-cylinder N-type engine with exhaust-induced tower cooling system. The single-speed transmission was very simple. The usual standard clutch pulley had a gear pinion attached which in turn drove a large diameter gear mounted on a countershaft. On the other end of the countershaft, on the opposite side of the tractor, a roller chain and sprockets drove the second countershaft, on the end of which was the final drive pinion meshing with an internal tooth-ring gear attached to the driving wheel. The driver was perched high up on a standard plough seat so that he could look over the top of the driving wheel and fuel tank, but even so, his vision was greatly impaired by the cooling tower in the front of the machine. Steering was by standard angled steering shaft, worm and wormwheel, and the front wheels were mounted on the usual stub axles. The three-wheel concept did not achieve popularity, and only two or three of these machines were built.

Competitor for Caterpillar?

Tracklaying vehicles were known in Australia by 1916, and McDonalds evidently decided that there were distinct advantages to be gained, so they built a prototype crawler version of the Steel Horse. Instead of the big driving wheel, it had a single track. The front-end arrangement was exactly the same as the updated wheeled version, with the clutch pulley mounted on the engine crankshaft, which in turn drove a set of reduction gears to a chain drive on the opposite side. The final drive was achieved by means of a pinion with teeth to match the rear track-driving sprocket. The pitch of the track chain appears to have been at least four inches, and the pads were red-gum wooden blocks bolted to a steel base attached to each chain link. The track had three lower idlers on spring mountings and three sets of tipper idlers. The construction of the track mechanism was advanced for its time. Both a wheeled and a crawler Steel Horse were exhibited at the Melbourne Royal Show in 1916, and one of the signs proclaimed the machine as 'British Juggernaut's baby brother ... the Steel Horse designed and built by A.H. McDonald & Co.' There is no evidence that McDonalds were using any patents or licences from overseas. Neither is there evidence to show whether more than one crawler Steel Horse was ever built.

Royal Show time was a special time for A.H., who liked to go out and meet clients at the company exhibit, which by 1918 had its own quaint hexagonal office at the Melbourne Showgrounds.

The Lightweight

The McDonald Lightweight was produced in late 1916. This four-wheeled tractor was powered by a single-cylinder N-type engine of 20 HP. The three-speed gearbox was fully enclosed and a differential was incorporated into the enclosed final drive gears from which the drive shaft extended across to pinions driving an internal-toothed ring gear on each drive wheel. The driver's compartment

The McDonald-built market prototype three-point undercarriage tractors called the Steel Horse in 1916. Below: 20 HP wheeled Steel Horse with single-cylinder N type engine.

The Crawler track version Steel Horse. The tricycle suspension system did not prove popular.

was improved, but the steering was still the old worm-driven shaft operating chains to the swivelling wagon-type axle.

The Lightweight tractor introduced a new look for McDonalds, when for the first time they used a modern type of water-tube radiator. It was set conventionally in the front of the machine, with a shrouded belt-driven fan at right angles to the radiator with ducts to direct the air stream through the radiator. Now an engine bonnet was fitted, extending from the radiator to the driver's cab, and a shorter canopy provided driver protection. Instead of sheet metal wheel guards, vertical tongue-and-grooved timber slats were placed on each side of the cab. Optional equipment included canvas blinds to drop down on each side of the engine for weather protection.

A McDonald Imperial tractor at work pulling a Massey-Harris No. 2 Reaper-Thresher near Maffra, Victoria, in 1914. Model EAA, with tower cooling system and 40 HP D-type engine. (Photo originally from Massey-Harris archives)

The N-type engine, despite being a single-cylinder of 8.5-inch bore and 9-inch stroke, was said to be 'surprisingly easy' to start. It had a McDonald-designed impulse magneto for a fat starting spark, and compression relief lever to hold the inlet valve open for part of the stroke. The flywheel rim was shaped to provide a good hand grip. With the carburettor primed and set right, gear in neutral, it only needed one pull on the flywheel to get started — if the engine was in good condition.

The EAA Tractor Series

The first three EAA tractors, introduced in 1914, used the 20 HP twin-cylinder D-type engine. This was the kerosene-fuelled spark-ignition side-valve engine with T-head, that is, with exhaust and inlet valves on opposite sides. The 1916 EAA tractors had a large capacity vertical-tube radiator cooling system. The radiator was mounted on springs and located fore–aft on the frame so that a belt-driven fan could be conveniently placed snug with the radiator core. The narrow front-end appearance of the EAA was quite distinguishable. The EAA that was put to work in the 1918 Werribee tractor trials was a later model equipped with twin-cylinder N-type L-head engine of 45 HP. The EAA series was produced from 1914 through to 1920.

The SD Lightweight Tractor

Around 1917 another small tractor, called the SD Lightweight, was designed by the McDonald team. This incorporated the single-cylinder N-type engine driving

Another view of a Model EAA tractor — the radiator was mounted fore–aft. This version was fan-cooled.

The 1916 Lightweight tractor with N-type single-cylinder vertical engine of 20 HP. This was the first time McDonalds adopted the conventional radiator and fan cooling layout, and they supplied an engine bonnet and driver canopy.

Advertisement in Land, Transport and Sea Trade, *1919 shows the McDonald SD Lightweight Imperial tractor with its vertical-cylinder N-type 20 HP engine. The short chassis and gooseneck up front was a convenience that enabled the tractor to make tight turns and it could be readily converted to road roller configuration at the factory.*

through the standard tractor or road-roller gearbox and differential mounted on the rear axle. The front part of the chassis was goosenecked to accommodate a central swivel pin and the front wheels were placed close together, pivoted on the centre of a short axle to accommodate rough ground. Steering was by worm and wormwheel operating on the central swivel pin, although some units had a chain arrangement, similar to early road rollers.

This gooseneck chassis arrangement was designed to allow the SD model to be converted into or manufactured as a road roller by simply removing the rear wheels and the front wheel and axle assembly, then replacing them with road rollers front and rear. Several machines were sold to municipal councils as convertibles to enable them to obtain the utmost use from the machines, either as a tractor working on road construction or as a roller to provide compaction. The SD had a three-ratio gearbox, giving road speeds of 1.5, 2.1 and 3.16 miles per hour forward and 0.7 in reverse. The roller version was called the Model SE.

As a farm tractor, the SD Lightweight was not popular because it was considered underpowered for its weight and could only pull light implements. Its main attribute was the small turning radius of 10 feet, as against the 20-foot swing of larger contemporary tractors.

The TB Tractor

About 1918 the engine, transmission and differential arrangement was updated and rationalised to suit a larger range of tractors and three-wheel style road rollers, in the weight range from 10 tons upwards. Some of the tractors were sold as convertible road rollers.

In the tractor form this machine was called the TB model and in the road roller form the TC. It was well thought through from a production point of view. The basic chassis, containing the power unit and transmission train right back to the rear axle and differential, could be produced on an assembly-line basis. It was comparatively simple then to finish the machine either as a tractor or a road roller, and in both versions it sold well — until phased out around 1923.

Werribee Tractor Trials

The first comparative tractor ploughing trials held in Australia took place on the State Research Farm, Werribee, in 1918. At the 1918 trials, McDonalds placed their 45 HP Model EAA and a Lightweight, both with N-type engines. These tractors were tested against the US-built Mogul, Bates' Steel Mule and Waterloo Boy, the Australian Jelbart and the Sunshine Models O and A.

In the following year McDonald's entries were the EAA, EAB (a modified EAA) and an SD Lightweight. The independent ploughing reports prudently indicated satisfactory performance from all tractors. Despite the faint praise, the McDonalds were pleased enough with the 1919 results to freely quote them in their promotional literature.

The EAB Tractor

The EAB for 1919 had a further-modified cooling system, with the radiator set conventionally across the chassis in front, but with the fan still situated sideways.

This was the last of the series, which was phased out in the early 1920s along with the McDonald spark-ignition engines, which were replaced by the Super Diesels.

McDonald Super Diesel Engines

The company started on diesel engines around 1917 with the design of their Super Diesel range, intended to supersede their spark-ignited engines. They found their niche in the engine market first with their C-type stationary diesels around 1920. It was patterned on an American design and Neal McDonald attests that even the name 'Super Diesel' was a US acquisition for these engines. They could operate on a range of fuels, including crude oil. The first diesel-powered McDonald tractor was a special, built for the Shire of Heytesbury, Victoria, for municipal road and quarry work. It was sold in 1922.

The McDonalds' first diesel-powered tractor was built for the Shire of Heytesbury, Victoria, for driving quarry plant. It was powered by the L-type 20 HP Super Diesel crude oil engine. This engine was similar in configuration to the McDonald Model K semi-diesel engine. Note the fuel heater wrapped around the exhaust pipe.

A 1932 advertisement pointed to the ability of the McDonald Super Diesel to run on cheap crude oil. (Power Farming, *November 1932)*

The company took the basic chassis of one of their larger tractors and stretched it a bit to accommodate the L-type 20 HP low-speed diesel engine. This engine was normally fitted with two spoked flywheels, but the tractor application could accommodate only one flywheel, so that the other end of the crankshaft could take the clutch pulley and gearbox drive. This single flywheel was fully five feet in diameter and weighed half a ton. A chain drive was used between crankshaft and gearbox because of the extended distance to the input transmission shaft. The gearbox with cover on top but exposed underneath was the current tractor/roller type, mounted under the engine.

The final drive used the usual external ring gear and differential hung on the rear axle inside the right-hand rear wheel. For quarry work it takes little imagination to consider how the dust from a stone crusher would collect on the tacky exposed gears and bevels — a combination that would make an ideal grinding paste!

The chassis was similar to the SD model but beefed up using main members of channel steel 15 inches deep. Steering was roller-style with open-worm and worm-wheel shaft acting on two chains to swivel the front axle which was mounted on a central kingpin. The engine was cooled by a fore–aft mounted radiator with friction-driven fan driven from the flywheel rim. Air starting was provided for the big single-cylinder diesel which would otherwise have been quite difficult to start manually.

An Era of Imports

After World War I, imported tractors, such as the Fordson, McCormick-Deering, and a number of others came onto the Australian market in increasing quantities to meet the demand. These imported tractors were powered by compact multi-cylinder vertical spark-ignition engines which operated at higher rpm, enabling the use of much smaller low-torque transmissions, so the tractors were not only more powerful but also more compact than the traditional McDonald machines, which had top engine speeds limited to around 500 rpm.

McDonalds were being placed increasingly out on a limb by this technology. Their heavily built low-speed engines ran at one half to one third the speed and were costly. The argument advanced for the low-speed engines was that if one engine ran at double the speed of the other, then it would have only half the life. That line of salesmanship was soon overtaken by events. Nonetheless, the company had been taking stock of the situation and decided for the time being to enter into a distribution arrangement with certain overseas tractor makers. By 1923, the last of the early series McDonald agricultural tractors was sold.

Bert Peakman, works foreman, was sent to Rockford, Illinois, to carry out final negotiations with the Emmerson-Brantingham Company (known generally by their initials E-B). He was an Englishman, with the attitude that nobody in the world could build any kind of machinery as well as the British, so it was a wonder that he was able to curb his feelings towards the Americans. He evidently must have compromised, as a distributorship contract was confirmed and initial orders were placed.

The Imperial E-B tractor was the Emerson-Brantingham from the US, marketed by the McDonalds between 1923 and 1927. (Power Farming, *May 1926)*

The first E-B tractors were imported in 1923. The US manufacturer had several models available, including the sleek Big 4 series of 20–35 HP and the Big 6 of 60 HP, but McDonalds chose to bring in only the 12–20 Model L, which they claimed 'had the power of a 1530 tractor for a price lower than that of scores of 12–20 tractors, thus saving £100 to £200'. These steel-wheeled tractors sold quite well and proved a success in the field.

In 1926, the E-B kerosene-powered 15–25 Model K became available, and it sold reasonably well for just under £500. Bear in mind, however, that by this time the Fordson was available for just £180.

When spare parts became a problem McDonalds made them at the factory on demand. In 1926 a vaporiser was designed locally to cope with problems of heavier fuel. Extension steering equipment and a power take-off were also developed. One tractor at least was fitted with special mud wheels.

Avance Tractor

This Swedish tractor was first imported by the company early in 1927, while the McDonalds were still selling the American E-B tractors. At a field day at Dookie Agricultural College that year the company displayed both makes of tractors. It had wanted to market a diesel-type tractor of the higher speed type in order to have an edge over the Fordson, and since E-B did not offer one, they turned to the Swedish design.

The sales team must have been fairly successful in forward selling the Avance tractor. Between the last week of April and the end of June 1927, 20 tractors were delivered, mostly to clients in Victoria and New South Wales. Two even went to South Australia and one to Western Australia. These sales accounted for most of the first shipment. The tractor with the lowest serial number was apparently used for demonstrations, as it was not sold until December 1927.

The 1928 model Avance Crude Oil Tractor was a refined version with a four-bearing crankshaft. The first models of this Swedish-built tractor marketed by McDonalds were equipped with a three-bearing crankshaft which was notoriously inclined to fail in service in Australia. The Swedish engine was a two-cylinder hot-bulb two-stroke. (Power Farming, *February 1928)*

The Avance was powered by a twin-cylinder, vertical two-cycle hot .plug engine, rated at 30–35 HP. The ignition plugs had to be heated for starting, and two separate systems were provided. An electrical heating system, driven by battery and generator, was the manufacturer's preferred system. The backup system involved heating the plugs by blowlamp until they were red hot, then an air starter system was used. The air receiver was recharged by using one cylinder as a compressor, while running the engine on the other. Ben Bingham from Western Australia, who serviced many Avance tractors at the time, said that although the electrical heating system worked reasonably well, Australian operators generally preferred the blow lamp. The hot plugs could be heated just about as quickly, but the engine seemed to start more readily because the cylinder head around the hot plug area was warmed up; besides, the electric heating system was difficult to clean.

The cast cylinders were of metric size, 180 mm bore. McDonalds must have been disappointed that they did not have renewable cylinder liners (a feature they had already adopted in their larger Super Diesels) because when the Avance bores were badly worn, the tractor had to be split where the engine joined the gearbox, then the engine dismantled and the cylinder block sent back to the works for reboring and fitting new pistons. Further fitting was required to big ends on reassembly. The whole job must have been very expensive for the tractor owner.

The Avance tractor design, while generally sound, did have a disastrous weakness — crankshaft failure — which showed up within a few months after use of the first tractors. There was an epidemic of broken crankshafts, in some cases causing damage to cylinder blocks which were smashed beyond repair. McDonalds were caught between angry clients demanding urgent replacements under warranty, and the manufacturers in Sweden. The McDonalds were not prepared for this disappointing turn of events, and had no spare crankshafts. In any case, fitting new crankshafts of the same design would not have prevented a recurrence of the problem. The cause of the problem was the use of a three-bearing design, which had inadequate support between the rear main and the flywheel, which, with the clutch assembly, weighed over 300 pounds.

Apparently the Swedes were not aware of the weakness, as it had not occurred in Europe. McDonald serviceman Ben Bingham attributed the problem to the use of a somewhat heavier fuel in Australia, which required a change in injection timing. Anyway, after a flurry of telegrams the Swedes were stirred into action. The Avance Company sent out an expert to work with McDonald's men to effect modifications. These included a four-bearing crankshaft, and a new cast housing to better take the weight of the flywheel. This housing added about six inches to the length of the tractor. A tuning modification kit and improved fan drive were also provided. Bingham and Anderson, the Swedish expert, worked for six weeks in the Balranald-Hay area converting tractors before moving down into Victoria. All 20 tractors in the first shipment were modified. Despite this, Ben Bingham had a healthy respect for the Avance tractor, regarding it as one of the best performers available at the time.

The original demonstration tractor was fitted with a new crankshaft before it was sold, in December 1927. To illustrate the short life some of them had before crankshaft breakage, one client, W.J. Collins of Beulah, Victoria, purchased his first Avance in May and by August it had failed and had to have the engine modified.

After several years' experience, McDonalds had had enough of other people's problems. They discontinued the Avance line early in 1930, when they were ready to introduce their own Imperial Super Diesel tractor.

Nevertheless, McDonalds did achieve success marketing the Avance marine engine, which had a good power-to-weight ratio and was well suited to fishing boats and launches.

Super Diesel Tractor Models TW and TWB

With the launch of their T-type diesel engine in 1929, McDonalds resumed local tractor production after a break of almost a decade. The Lanz Bulldog engine design influence was evident in their T series, but not the Lanz transmission. The TX tractor, first sold in 1930, was equipped with the T-type engine mounted in an imported Rumely chassis purchased under an agreement with the American company, which was using a similar engine layout, but spark-ignited. As McDonalds tooled up to produce their own chassis design, the model designation was changed to the TW in 1931.

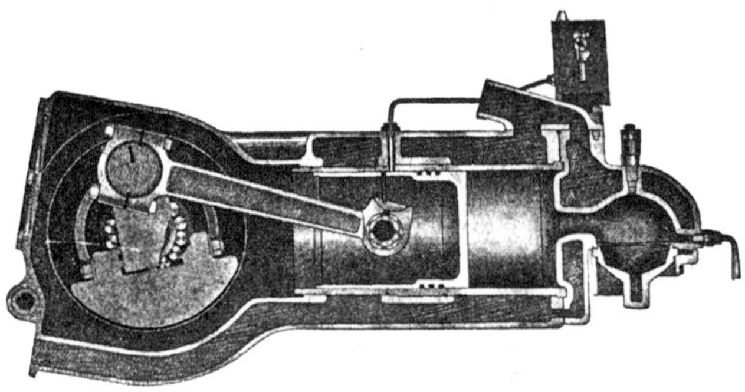

The McDonald Imperial Super Diesel engine reflected the German Lanz tractor engine influence, with its hot-bulb cylinder head design. The 30–36 HP engine, of 9-inch bore x 10-inch stroke, was called the T Type. The engine had a wide-range fuel tolerance.

Ray Gillett, of Ringwood, Victoria, a specialist on diesel engines in Australia, suggests that it is more than a coincidence that the takeover of Rumely by Allis Chalmers in 1929 may have 'orphaned' some Rumely tractors that were in dealers' hands at the time. In Melbourne, the Rumely dealer was Waugh and Josephson, which would be anxious to offload stock. The Rumely Model X tractors were apparently acquired complete but the Rumely engine was sold for stationary applications and the McDonald diesel put in its place. The TW tractor was built around the slightly smaller Model W Rumely. The Model TWB, which came out in 1932, had a McDonald-built chassis and T engine but Rumely transmission. As Rumely spares became less accessible and the company had the time to tool up, McDonalds gradually assumed complete fabrication on their own (R. Gillett, personal communications).

The Great Depression inevitably hurt the McDonald tractor business. After struggling to sell just four of the test model TX tractors in 1930, then 10 TW tractors in 1931, they had to face severely delayed payments and extend terms to hard-pressed farmers. If there was a poor harvest at the time, there was little alternative but to extend terms. Bob Clark, the first specialist salesman to be employed by McDonalds, had plenty of problems selling tractors in the 1930s. On one occasion he telephoned the boss to report a sale. The conversation went something like this:

'Good morning A.H., I have booked a tractor order.'

'That's excellent, keep up the good work.'

'There is just one catch, A.H., in order to clinch this deal I had to trade in a team of 12 draughthorses.'

'What did you say? — Horses! That's no good, Bob, what the hell are we going to do with a team of horses?' A.H. fairly exploded.

'I don't know, A.H., but that's the only way I could get the order and the opposition agreed to trade them, so we had to do the same.'

'Look Bob, I don't know what we can do with 12 horses, have you got any suggestions?'

'I did hear of somebody down the line that might be interested in them.'

'Well you'd better get down there and find a buyer for them right away, otherwise we'll all be in trouble.'

Needless to say, with his livelihood at stake, Bob moved with some alacrity to dispose of the horses. Each and every tractor sale was vital.

The TWB was the model designation in 1932 when 15 tractors were sold, 32 in 1933, 50 in 1934, and, as the Imperial's reputation grew, 60 in 1935. Two Model TWBs were exported to India in 1933 (Numbers 98 TWB and 123 TWB). In 1934 a distributor was appointed for New Zealand. The first tractor shipped there (No. 226 TWB) was well accepted and many more followed until the disruption of World War II.

An interesting point about the later series tractors was that while they started out with the Lanz tractor configuration, they were changed so that the engine cylinder head was at the rear and the frame reworked to put the engine further rearward to take some weight off the front axle for better steering.

Around 600 of the 30 HP TWB McDonald Imperial tractors were made between 1932 and 1944. Rubber tyres were fitted by McDonalds in 1938. The Rumely influence was still apparent in the tractor's styling. Fourteen TWBs were shipped to Perth in 1937, which was a good year for the WA branch.

Steel Wheels to Pneumatics

McDonald's installed special angled lugs on their steel wheels for the rice country of New South Wales. With the acceptance of pneumatic tyres, beginning in Australia around 1934, the company at first cut back their standard wheel hubs and spokes to attach steel rims to take pneumatic tyres. Within a short period, both Dunlop and Goodyear began offering fully cast wheels to suit the McDonald tractors. Both these rubber companies had to struggle hard initially to win the farmers' confidence for pneumatics, and entering the wheel and rim business was part of the strategy they adopted.

By 1938, however, most tractors were being offered with pneumatic tyres. For the McDonalds this required design modifications in the brakes and altered gear ratios for higher travel speeds. The power take-off was added as standard in 1938.

By 1936 production was up to one tractor a day. Four hundred McDonald Imperials were made between 1936 and 1938. About 600 of the TWBs were sold by 1941, when production was switched to wartime activities.

McDonald's Perth branch was operated by a very able manager, Andrew Fisher, who built up sales for their best year in 1937. Film footage of 14 tractors proceeding through the streets of Perth for loading onto rail for delivery to far-flung Western Australian farmers is preserved in the National Archives in Canberra.

Post-war Activity

After World War II, an upgraded TL engine of 40 HP rating, but with same bore and stroke, was adopted for the McDonald tractor, which was built around a modified TWB tractor (called the T6A). This tractor was rushed together for the 1946 Melbourne Show. But the pto clutch did not work well, so it was redesigned for the model T6B tractors, eight of which were sold between 1947 and 1949. Still, the wet-plate pto clutch was unsatisfactory. A dry-type clutch was adopted for the Model T6C, launched in 1949. Ten tractors were sold and plans made to build 100 more, as there was a backlog of orders for tractor-hungry post-war agriculturalists. But some parts, such as ball bearings, were practically unobtainable. There were competing demands on the factory for production of road rollers. Such was the confusion that tractor orders suffered prolonged postponement due to the inability of the factory to cope. After several years of frustration, management abandoned farm tractor production to concentrate on their other product lines. The last McDonald tractor delivered was a T6C, No. 7032, in 1955.

The company was taken over by Jacques Ltd and that company maintained production of only the road roller under the Imperial logo. This was at the Jacques-McDonald Division of the company, in Clayton, Victoria.

After World War II McDonalds produced the T6A, T6B and T6C tractors, still with the single-cylinder hot-bulb Imperial Super Diesel engine, until production petered out in 1955. The loop-scavenged two-stroke engine was rated 40 HP at 500 rpm.

[I am extremely grateful to Neal McDonald, son of A.H., who dedicated his professional career to working with the company. Neal McDonald has prepared a biography, *A.H. McDonald: Industrial Pioneer* (1988). The photos in this chapter are from Neal McDonald's collection, unless otherwise indicated.]

McDonald Tractors at a Glance

1908–11	EA Type tractors (first one sold in 1909).
1911	Austral-Otis special
1911	Irrigation Ditcher Special
1911–16	The company collaborated with Frank Bottrill to build tractors incorporating Bottrill's Dreadnaught wheels
1911–14	EB Type Imperial
1912–15	The EC Type
1913	The ED Type
1914–20	The EAA Type
1916	The Steel Horse series (including one crawler track version)
1919	EAB model
1916–18	The Lightweight
1918–23	The TB
1922	Y-Type Special for roadworks
1923–27	Emerson-Brantingham (Racine-built imports)
1927–30	Avance (Swedish-built imports)
1930–44	TX, TW, TWB. McDonalds resume local design and production of 30 HP semi-diesel engines for their tractors and equipment
1935	TWV Special for roadworks (replaced Y-Type) [World War II intervenes]
1946	Prototype of the 40 HP T6 Series built — the T6A
1947–49	The T6B, six built
1955	End of McDonald tractor production. The number of McDonald tractors made was actually less than 1000 in total

10 CALDWELL-VALE — AUSTRALIA'S FIRST FOUR-WHEEL-DRIVE TRACTOR

Like the horse, the weight of the machine must be borne by the soil at four points of contact; therefore, to attain the minimum of weight with the maximum of power, the tractor should be driven through each wheel — rear wheels must track with the front or steering wheels so as to offer the least resistance of the machine over soft earth. With a four-wheel-drive the machine would be a true tractor, and its action analogous to that of the horse.

— *Caldwell-Vale company literature, c. 1913*

The first Caldwell Brothers 60/80 HP road train hauling a gross load of 50 tons in South Australia, c. 1910.

The first Australian four-wheel-drive I-C tractor was the brainchild of brothers Felix and Norman Caldwell of Adelaide, South Australia (Giltrap, 1951). The brothers patented their ideas 'for transmitting power equally to all four wheels of motor-driven vehicles' in 1907.

To test the principle they first constructed a 'motor plough'. They mid-mounted a 10-plate disc plough under the belly of the tractor. The first tests of this machine took place at Roseworthy Agricultural College, South Australia. Its principal (the same Professor Perkins who was a strong advocate of horse farming and earlier opposed 'tractorisation' on economic grounds) reported favourably on the 10-disc motor plough, which was claimed to be capable of two acres an hour, ploughing a width of six feet to a depth of eight inches. It was also successfully demonstrated on sand at Henley Beach, South Australia. This tractor was powered by a 20 HP petrol engine and weighed six tons. It had power steering, so that, according to the 1913 Caldwell-Vale company booklet, 'even a boy can with ease manipulate the largest tractors cultivating from eight to ten acres per hour'.

The Caldwell brothers next built and demonstrated a 60/80 Road Tractor. An Adelaide merchant was so impressed with the first model that he bought the four-wheel-drive machine on the spot. He used it to haul trailer loads of up to 50 tons into the backblocks of South Australia.

Sometime during 1910 the brothers teamed up with Henry Vale, a financier and heavy equipment manufacturer who was known as a prominent member of the fledgling New South Wales motoring community. The Caldwell-Vale Motor and Tractor Construction Co. Ltd was subsequently established in Auburn, a western suburb of Sydney. Demand for their four-wheel-drive 'motor tractors' was such that the factory was soon expanded to enable production of 40/50 HP and 60/80 HP machines. In 1913 a four-wheel-drive, four-wheel-steer car was also built. The 30 HP motor vehicle was demonstrated in May 1913. It was the only one produced.

Motor Plough with 30 HP water-cooled petrol engine; the 10 discs were mid-mounted and had a power lift system. Six tons all-up weight. The four drive-wheels were each of 4 foot 6 inch diameter and 14 inches wide. The front pair was used for steering.

Company Downfall

The company faced a major distraction with a lawsuit brought by one of its clients concerning the guaranteed performance of its trucks. The company lost the suit and

Caldwell-Vale automobile: four-wheel-drive, four-wheel-steer, 30 HP, four-speed gearbox.

by 1916 the Auburn works were in the hands of Thomas Purcell, Director of Purcell Engineering Co. Ltd, Auburn. It is possible Purcell was the litigant.

Thomas Purcell had a subsidiary haulage company in Longreach, Queensland. There he used a number of the Caldwell-Vale road tractors for carting wool and other goods to and from far-flung stations.

During their brief span, the innovative Caldwell-Vale team generated a diverse product line which included their tractors — agricultural and road-type (around 50 may have been built) — the four-wheel-drive automobile, light-rail locomotives (from 30 to 140 HP) and their own range of stationary engines.

60/80 HP Caldwell-Vale motor tractor under test hauling 20-ton load.

A later version Caldwell-Vale 4WD tractor doing paddock work. (Side-and rear view photos courtesy Bruce McDonald, other illustrations derived from a company bulletin, Motor-Traction, *c. 1913.)*

The NSW Caldwell-Vale works in 1913.

e first commercial Australian-built tractor, an Imperial Oil Tractor Model EA by A.H. McDonald & Co., was exhibited at the Royal Melbourne Show in 1909. It erated on kerosene. (Courtesy The Old Machinery Magazine, *Feb/Mar 2005).*

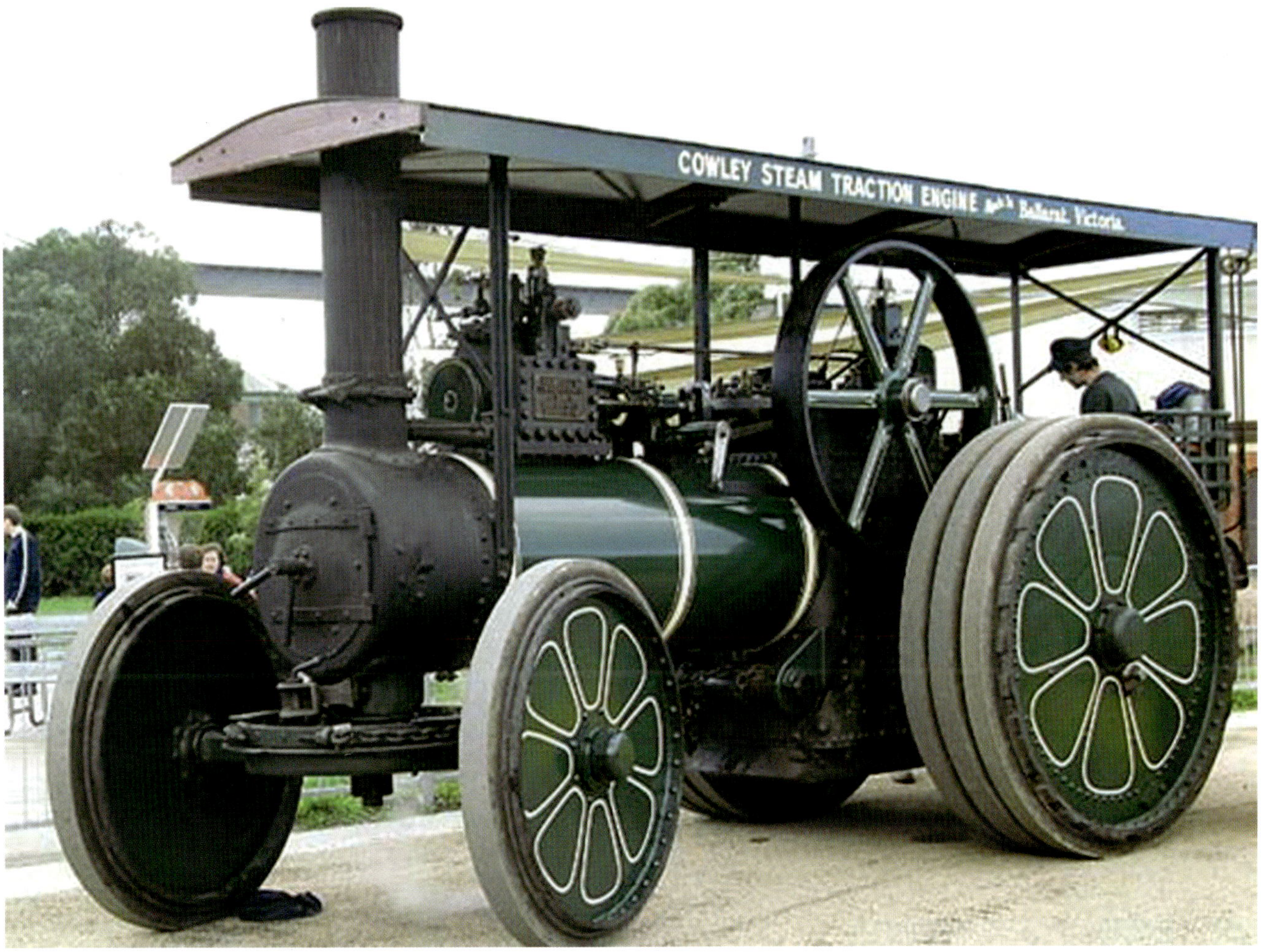

Cowley Steam Traction engine made by Cowley's Eureka Iron Works, Ballarat, around 1913. This was one of the last steamers made in Australia. (Illustration courtesy Martin Hallett, Museum of Victoria)

Norm McKenzie of Cumnock, NSW, owns this 1903 Ivel tractor. There are several Ivel tractors in Australia. Norm McKenzies' is serial number 269. It is one the oldest operable tractors in the world. (Photo courtesy Val McKenzie)

In April 2003, a total of 1914 tractors was mustered on a farm near Cootamundra, NSW, establishing a world record for the greatest number of tractors operating at the same time in one place. The Guinness Book of Records *has acknowledged this as the largest gathering of tractors at work. Actually 13 tractors did not make it around the circuit so the recorded number is 1901. McKenzies' Ivel was in this group. Some 23 charities benefited from the 2004 assemblage. (Photo courtesy Vic Muscat)*

Reg Schuster of Palm Beach, Queensland, has restored a Caldwell-Vale motor truck. Here it is after restoration. Reg shows it regularly at the South Coast Restoration Society's Annual Rally. Inset shows the truck as found. The Caldwell-Vale company built Australia's first four-wheel-drive tractor similar to this truck. (Photos supplied by Reg Schuster)

1917 Jelbart owned by Jack Burrell of Gilgandra, NSW

The Ronaldson-Tippet 20-36 SuperDrive Tractor on rubber in 1937. (Courtesy** The Old Machinery Magazine, **August/September, 2004)

11 BALLARAT'S DEUTSCHER AND JELBART TRACTORS

I've heard young engineers in my own shop complain that they've arrived too late to make any discoveries. That's all nonsense. All inventors stand on the shoulders of the inventors who have gone before them, and the bigger the inventors, the higher the newcomers can stand.

— *R.G. Le Tourneau*

Deutscher, Lutz and Gem

In 1908, the Deutscher family and partners commenced making their Gem benzine-and-oil single-cylinder horizontal engines at their Clyde Engineering Works in Hamilton, Victoria. Between 1908 and 1916 they also built a vertical Little Gem motor. Their engines ranged from 2 to 20 HP.

Around 1912 the company built at least two tractors. In the company booklet they were called 'Benzine and Oil Traction Engines'. This name reflects steam-era thinking, particularly as a cable winding drum was also offered: 'No tractor is complete without a good winding drum'. These tractors were used on properties in the district. They were equipped with Gem horizontal two-cylinder, water-cooled engines.

According to Cliff Brinkmann, of Cavendish, Victoria, a specialist on and collector of Deutscher memorabilia, the tractor project was a dismal failure and almost sent the company broke.

Two brothers who worked as pattern makers at the Deutscher foundry in Hamilton were Oscar and Sonny Lutz, who originally hailed from South Australia. They believed they could improve on the Deutscher designs so they left and set up an independent engine business in Melbourne. Around 1914 they advertised the Lutz Farm Tractor in their product catalogue, as a tractor for which 'it is not necessary to provide special implements ... the implements ordinarily drawn by horses can be used with perfect economy and satisfaction'.

It is not known how many Lutz tractors were built — or if any were sold at all. As for Lutz engines, these were made at the Melbourne plant for about five years, between 1912 and around 1917.

Gem engines were made in Hamilton from 1908 to 1917, when the enterprise was known as Deutscher Son & Walker. Between 1917 and 1922 the company was known as C.T. Deutscher & Son. In 1922 Deutschers sold out in Hamilton and moved to Ballarat, where a branch of the family went on to make ride-on mowers, post borers and other farm equipment and continue today.

[Source material for this segment was provided by Cliff Brinkmann, Toolka, Cavendish, Victoria. The tractor illustration was drawn from a copy of the Deutscher company catalogue of 1912.]

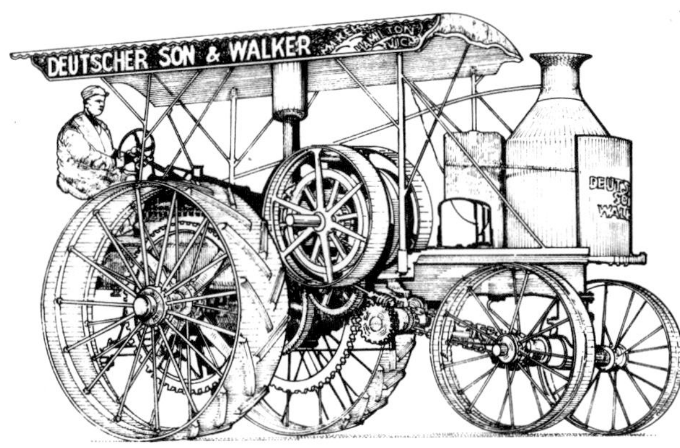

1912 Deutscher Gem 'Benzine and Oil Traction Engine'. It is probable that the machine was more suited to belt work or, at the most, only light pulling, based on a scrutiny of the drive wheels.

The Jelbart Bros

Ralph Sadlier Falkiner, entrepreneur, inventor, and wealthy manager of vast farming estates, was a shadowy but important figure behind at least two Australian tractor enterprises. He gave financial support to Frank Bottrill, and he was a shareholder in the company founded by the Jelbarts in Ballarat.

The Jelbart story begins with the arrival in Australia on the *Zenobia* of 21-year-old Thomas Jelbart from Cornwall in 1869. He was a blacksmith and farmer. Later, when his sons came along, he participated with them in developing machinery for Australian farming. The Jelbart Firm was in operation before 1911, and it was formally registered as Jelbart Bros by George E. and Frank S. Jelbart in January of that year. It operated from its Dana Street factory in Ballarat.

First Australian-designed Internal Combustion Engine?

The brothers produced the Jelbart Air Scavenge engine, patented first by F.S. Jelbart in 1909 but in all likelihood assembled some years earlier (perhaps 1904, according to Martin Hallett at the Museum of Victoria). This engine was a two-stroke, spark-ignited, single-cylinder unit, the cylinder head design reflecting the 1900 Ricardo influence from England. In 1912 it was available in four models: the A-type at £125 for an 8 HP portable, or £115 for an 8 HP stationary; and the B-type at £255 for a 12½ HP portable, or £245 for a 12½ HP stationary.

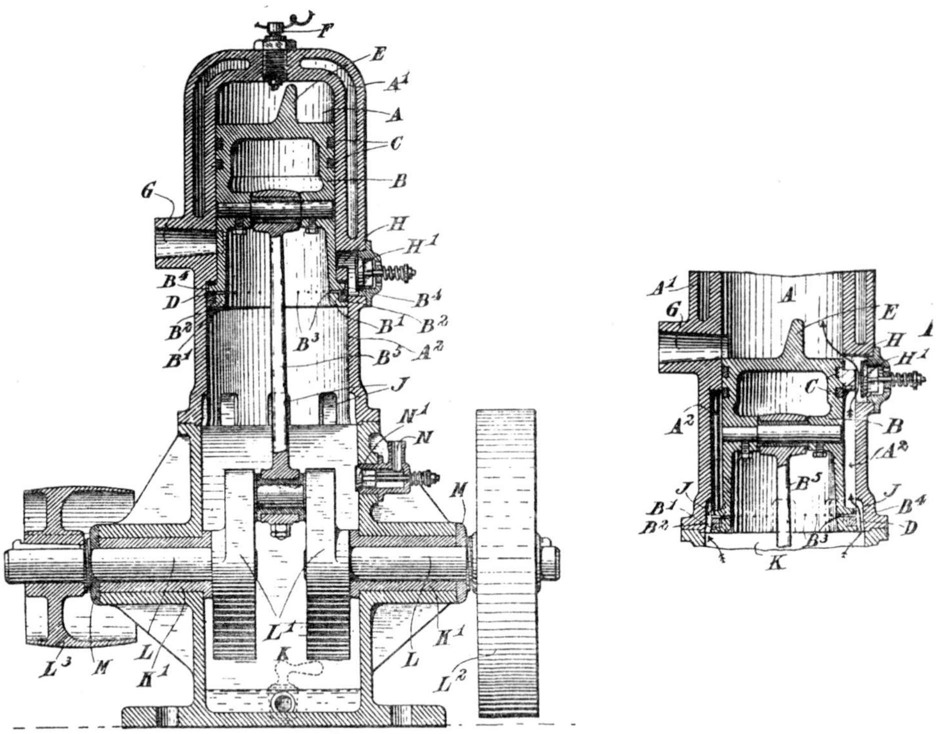

The Jelbart single-cylinder two-stroke engine design, as disclosed in Australian patent No. 15,1741/09, 21 July 1909 to F.S. Jelbart. The Jelbarts were among Australia's first internal combustion engine designers and manufacturers. Note the skirted piston, used to create suction, then compression in the crankcase, first drawing in air, then forcing the air/fuel mixture into the combustion chamber as the cylinder ports were uncovered. Jelbart literature boasted that their first engine was still working in 1933.

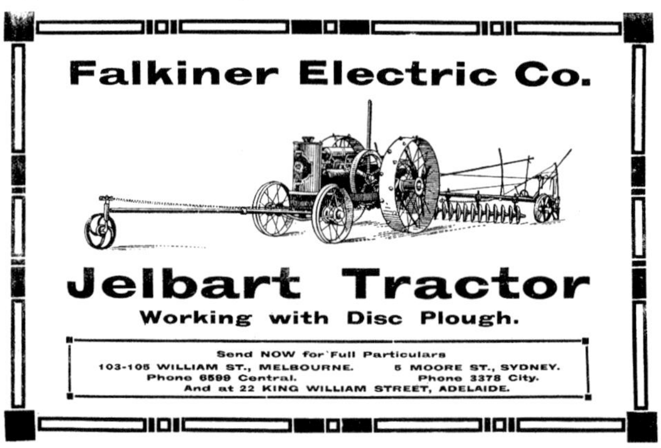

'Jelbart tractor working with disc plough' in a 1916 advertisement for the Falkiner Electric Co. The tractor was equipped with 5 foot 6 inch drive wheels with final open pinion and segmented bull gear on the wheel spokes.

This 1917 Jelbart illustration clearly shows the segmented belt drive to the transmission. The advertisement appeared for several years in the Land and Transport *journal up to the 1920s. (Photo courtesy* The Old Machinery Magazine*)*

On all four models the cylinders were water cooled, except for the area around the cylinder cap, known as the 'hothead.' These hothead engines were not so easy to start for the uninitiated. In the early days there were few experienced or competent engine operators around, so the Jelbarts were prepared to send an expert free of charge for four days to start up and instruct an owner about the operation and maintenance of a new engine. Jelbart engines were started on benzine or petrol, then, when warm, switched over to the cheaper fuel, kerosene. In 1914 or thereabouts, an engine unit was fitted to a tractor. The first Jelbart tractors used their No. 8 oil engine, with 7-inch bore x 9-inch stroke and piston flared to nine inches diameter at the skirt. It developed 14 HP at 400 rpm and was accordingly rated the equivalent of eight horses.

In 1915 R.S. Falkiner, owner of famed Boonoke station and Australia's largest merino wool producer at the time, was listed as a major shareholder (holding around 50%). The company was registered in February 1916 as Jelbart's Proprietary Limited with the Falkiner Electric Co. as agent for its tractors in Melbourne, Sydney and Adelaide.

In its heyday — which apparently lasted only until around 1926 — the company manufactured several sizes of engines, tractors, chaffcutters, gas producers and road rollers.

The Falkiner influence was also manifested around 1920, when the company built a prototype sugarcane harvester (mechanising the cane harvest was an intense personal interest of R.S. Falkiner). Even though the H.V. McKay & Massey Harris people inspected this machine, it was apparently not pursued through to production.

In 1918 Frank Jelbart patented an improved

Jelbart tractor ploughing in New Zealand. The steering assist was practically essential, both to ease the heavy work of steering and perhaps to aid the driver who could hardly see where he was going with this 'low profile' canopy. (Drawn from an illustration in Land and Transport, *1917)*

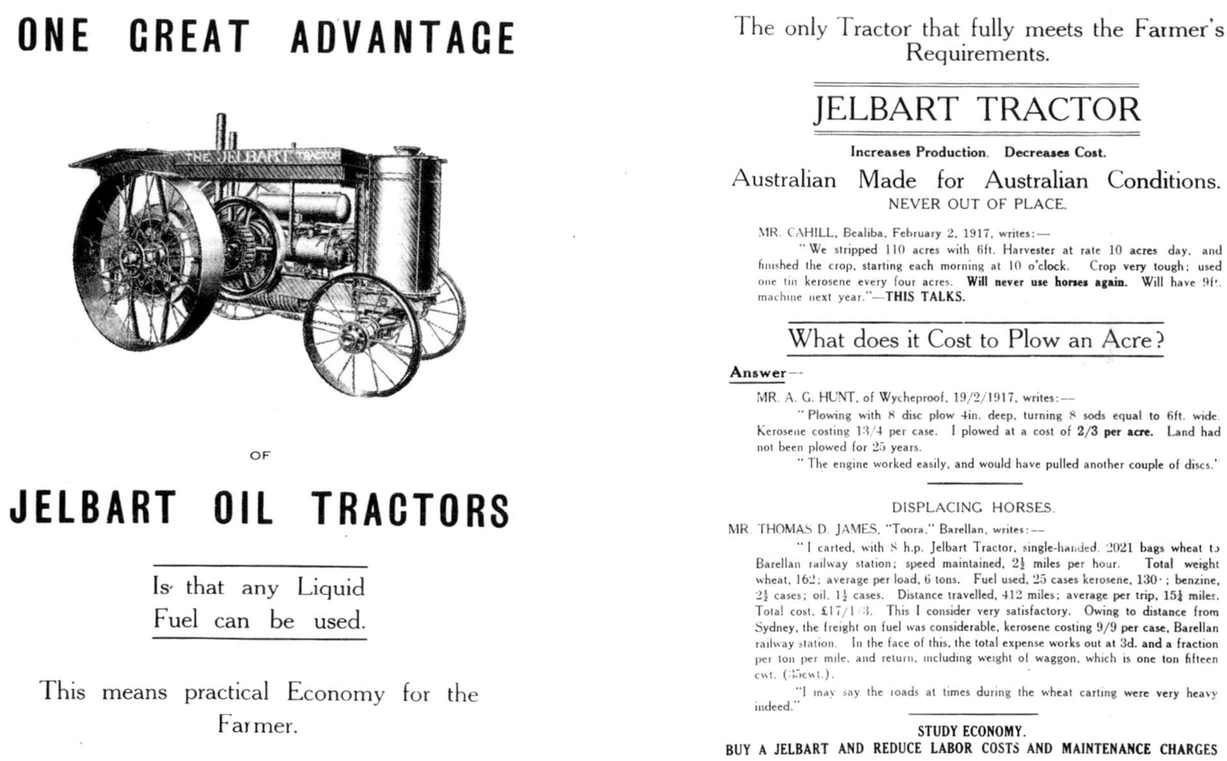
ONE GREAT ADVANTAGE

OF

JELBART OIL TRACTORS

Is that any Liquid Fuel can be used.

This means practical Economy for the Farmer.

The only Tractor that fully meets the Farmer's Requirements.

JELBART TRACTOR

Increases Production. Decreases Cost.

Australian Made for Australian Conditions.

NEVER OUT OF PLACE.

MR. CAHILL, Bealiba, February 2, 1917, writes:—

"We stripped 110 acres with 6ft. Harvester at rate 10 **acres** day, and finished the crop, starting each morning at 10 o'clock. Crop **very tough**; used one tin kerosene every four acres. **Will never use horses again.** Will have 9ft. machine next year."—**THIS TALKS.**

What does it Cost to Plow an Acre?

Answer—

MR. A. G. HUNT, of Wycheproof, 19/2/1917, writes:—

"Plowing with 8 disc plow 4in. deep, turning 8 sods equal to 6ft. wide. Kerosene costing 13/4 per case. I plowed at a cost of **2/3 per acre.** Land had not been plowed for 25 years.

"The engine worked easily, and would have pulled another couple of discs."

DISPLACING HORSES.

MR. THOMAS D. JAMES, "Toora," Barellan, writes:—

"I carted, with 8 h.p. Jelbart Tractor, single-handed, 2021 bags wheat to Barellan railway station; speed maintained, 2½ miles per hour. Total weight wheat, 162; average per load, 6 tons. Fuel used, 25 cases kerosene, 130·; benzine, 2½ cases; oil, 1½ cases. Distance travelled, 412 miles; average per trip, 15¾ miles. Total cost, £17/1/3. This I consider very satisfactory. Owing to distance from Sydney, the freight on fuel was considerable, kerosene costing 9/9 per case, Barellan railway station. In the face of this, the total expense works out at 3d. and a fraction per ton per mile, and return, including weight of waggon, which is one ton fifteen cwt. (35cwt.).

"I may say the roads at times during the wheat carting were very heavy indeed."

STUDY ECONOMY.

BUY A JELBART AND REDUCE LABOR COSTS AND MAINTENANCE CHARGES

In 1917, T. Robinson & Co. Proprietary Limited also handled the Jelbart tractor. This is an opening from their product catalogue, which included testimonials from satisfied Jelbart tractor owners. The Jelbart cabin was low and sleek, but hardly practicable.

engine design. This engine was also a two-stroke, with the uncooled hot spot in the cylinder head to aid in combustion of heavier fuels such as crude oil. The stepped-piston design was retained and there were ports in the cylinder block to admit air to the crankcase.

Jelbart advertising made much of the fact that their engines could run on 'crude oil'. Crude oil was said to be 'half the cost of kerosene' at 6½ pence per gallon in 1924. The engine still started on benzine, then ran on kerosene or crude oil when the vaporisor was hot enough, that is, after 5–10 minutes.

Role of Water in the Combustion Process

There was an interesting description of the water-injection knock control device with the Jelbart. Here is how their handbook instructed the operator to use the water nozzles, which could be controlled from the tractor seat: 'WATER SPRAY: After the engine has been working on crude oil or (especially) kerosene for a few minutes when on a fair load, you will hear a sound like the knocking of metal. This is only a gas knock, and is caused by highly heated gas due to high compression. Reduce this sound by turning on the Water Adjuster to the required amount. (A slightly pinging sound should remain.)'

Manufacturers realised that by designing engines

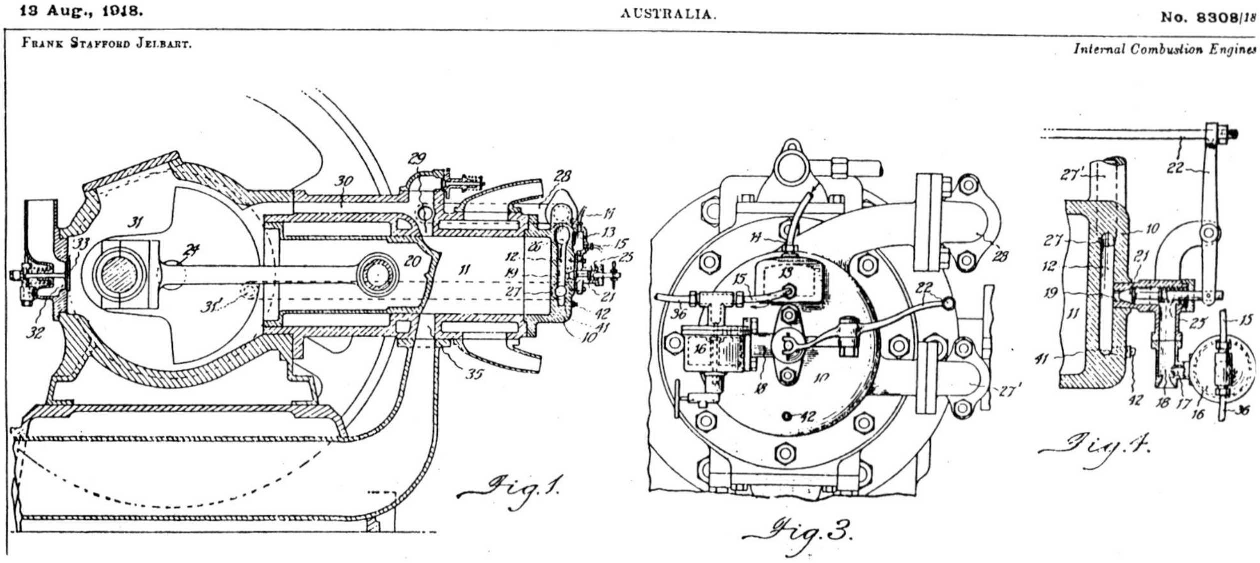

F.S. Jelbart's 1918 engine design patent. Below right: actual piston.

with somewhat higher compression, performance could be markedly improved by injecting water into the vaporiser along with the fuel mixture. Water injection also reduced the buildup of carbon deposits, enabling engines to operate much longer between services. Jelbart's design was different in one respect: the water was injected direct into the cylinder head, and not into the vaporiser. Water could also be sprayed into the silencer of the Jelbart to reduce fire risk in hot weather. Both water sprays were fed from the 30 gallon cooling tank and were under governor control, with water injection only being possible under load.

Engine lubrication was by sight-feed drip oiler onto the piston, set to give 20 drops per minute under load. There was also daily oiling of bearings and of the chain drive to the Bosch magneto. The spark-ignition system was designed so that 'the timing of the ignition is fixed, and cannot be disarranged by inexpert handling'.

The tractor used a 5-inch wide chain-link leather-belt drive from the flywheel-mounted pulley to the three-speed and one reverse gearbox. This belt served as a clutch, being controlled by an idler pulley. The gearbox itself was unusual inasmuch as it employed a wooden lid for easy access and oiling. Four engines sizes, rated between 8 and 14 HP, were available for the customer's choice in the 1922 tractor.

By 1926 the company had fallen upon difficult times and was officially deregistered. It was later reformed in Ballarat several times — as a jobbing business but not a tractor producer. Jelbart parts were available until 1958, when the company finally shut down.

[I am indebted to Ron Maslin of Temora and to Martin Hallett of the Museum of Victoria for helpful information on Jelbart tractors.]

12 RONALDSON BROS AND TIPPETT, BALLARAT

'We have hitched our tractor to the stars?'
— *Company advertising in 1937*

Dave Ronaldson, 1874–1937. H. Jack Tippett, 1872–1957 (Photos from University of Melbourne Archives)

The Smeaton district of Victoria is the setting for the story of this Australian company. It is an area of grazing country consisting of volcanic soil overlying rich alluvial goldfields. Gold was still being mined at the time that the Ronaldson Brothers grew up on the family farm in the district. Dave Ronaldson studied engineering at the Ballarat School of Mines, then was apprenticed to his uncle, James Smith, founder of the engineering firm Jas Smith Pty Ltd. This company is still in business, making farm equipment at their original Creswick Road works, Ballarat.

Establishment of RBT

In 1903, at age 29, Dave Ronaldson struck out on his own to produce farm machinery. He was joined in the first year by Adam, his older brother. Adam L. Ronaldson (1859–1916) was experienced in mining both overseas and in Victoria. He had even operated a drilling rig in Russia before he joined his brother's establishment. According to their 1904/05 illustrated catalogue, horseworks, chaffcutters, corn crushers, bag and hay elevators — and engines — were being produced. The Ronaldsons were also agents for the British-built Aveling & Porter steam traction engines and road rollers.

Jack Tippett joined them in 1905, at the age of 33. He was farming with his brothers on the family property, Scrub Hill, Newlyn. His father had threshing rigs with which he used to travel the area on contract, starting the harvest season near Bendigo. In 1892, the Tippett family purchased one of the first British Fowler steam traction engines in the district, in order to expand their contract work. In 1904 Jack Tippett met Dave Ronaldson at Newlyn. Jack was looking for off-season employment and was permitted to work in the Ronaldsons' Ballarat factory to obtain experience, with the idea of establishing a workshop at Scrub Hill. Before long Tippett became

SINGLE CYLINDER

Portable Oil Engine.

This is similar in construction to the double cylinder but of only 3½ b.h.p., and is made for driving smaller chaffcutters, shearing machines, hay stacking machines, &c.

3½ HP vertical single-cylinder Austral portable engine (Ronaldson & Co. catalogue, 1904/05).

financially involved and was soon made co-director in the firm, which on 1 April 1905 became the Ronaldson Bros and Tippett partnership (RBT), at Creswick Rd, Ballarat.

The partners had noted that other companies were importing internal combustion engines for farming, so they decided to build their own engine. They started with the vertical configuration, making a 3 HP single and a 7 HP twin cylinder to Dave Ronaldson's design. These were among the earliest internal combustion engines locally designed and built in Australia. They would run 'on the cheapest Russian oil kerosene and could be started in 2½ minutes'.

Ronaldson machines were first exhibited at the Melbourne Royal Show in September 1905. The company used the same show stand for decades thereafter. The Austral engine series came next. These were horizontal-cylinder engines with hot-bulb ignition, surface tension (vaporiser) carburation (following a British Blackstone patent) and designed to run on kerosene. The 3 HP ran at 360 rpm and the 22 HP model at 210 rpm. The one hundredth engine was completed at the works in March 1908.

First Tractor Prototype

At some stage before December 1910 the first RBT tractor prototype was constructed around one of these horizontal-cylinder Austral engines. This tractor was not put into production at the time, as the company decided it was better to concentrate on farm machinery and engine production.

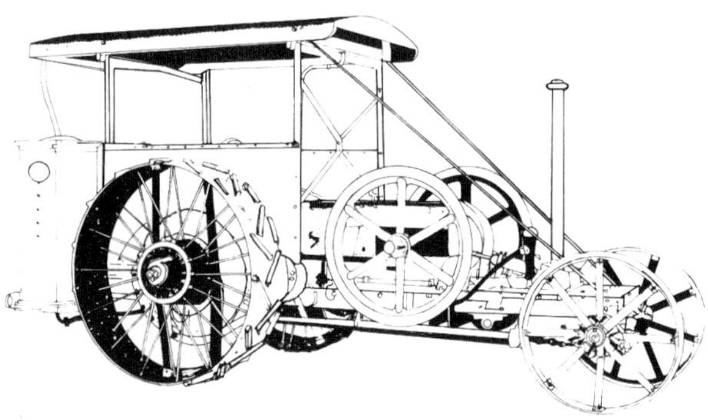

The prototype RBT tractor was assembled around a horizontal-cylinder engine. The 20 HP tractor weighed six tons.

The prototype RBT was displayed with 'Stacey's Traction Engine Treaders' fitted for better flotation — the treaders enabling the tractor 'to pass at half Power'. (Australasian Hardware and Machinery, *1 December 1910*)

Engine designer Dave Ronaldson was prepared to innovate. He once made an engine that ran on carbide, but it apparently lacked adequate control valves. After an incident in which a carbide-fuelled engine blew up and took a number of sheets of iron off the building around it, carbide as a fuel was not persevered with. For many years, however, the RBT factory and its overhead works were powered by another liquid fuel alternative: RBT engines running on suction gas. Stringybark wood was fed into gas producers and the gas was piped to the engines, which drove the works' overhead driveshafts and pulleys.

By 1923 the company had produced 4700 engines. In 1924 the firm exhibited machinery on the Australian stand at the Wembley Exhibition in England. About this time, too, the firm entered an agreement with Campbell's of Halifax, England, to make large single-cylinder diesels — the massive flywheels alone weighed three tons. The first unit was installed in the Ballan Electric Power Station in Victoria. Others went all over Australia for power generation and water pumping,

The Wisconsin Connection

In 1924 the firm began importing US-built Wisconsin engines. The directors were becoming increasingly concerned about the competition that their portable power plants were getting from tractors, so they conducted a survey to sample farmer opinions in the Victorian Mallee. They soon learned that the competition's new-fangled tractors were not only used for field work but were also being used as stationaries for chaffcutters and for feed milling, pumping, driving saw benches, etc. The assessment of the survey resulted in the decision to go into tractor production. The first serial production RBT tractor, the SuperDrive, came out in 1924. Initially the £700 tractor was built using many imported components, generally based on the contemporary US-built Illinois 18-30 SuperDrive tractor, but eventually most of the tractor was produced at the Ballarat works, except for the 30 HP water-cooled Wisconsin engine. Light oil was a cheaper fuel than kerosene or petrol around this time, so a special vaporising manifold was developed and patented to enable the SuperDrive tractor to be worked on this after warm-up.

Company advertising in 1928 cast aspersions on diesel engines; this was because some of the diesel or semi-diesel engines of the day were cantankerous to start, so much so that some farmers started them at the beginning of the cropping season and did not stop them until the end of the season! They might be left running overnight in the

Operator's approach to the Ronaldson-Tippett SuperDrive Tractor of 1927, with a four-cylinder Wisconsin spark-ignited petrol/ kerosene fuelled engine. Belt power 30 HP, drawbar power 18 HP, hand clutch to three-speed gearbox. Planetary gear final drive reduction through to the rear steel driving wheels.

A 1926 advertisement in Power Farming *for the 18-30 SuperDrive Tractor.*

Company literature showed a version of the SuperDrive fitted with scrub-rolling attachments. These included a bumper plate, radiator guard, transmission protection plate and sand-grips for the drive wheels.

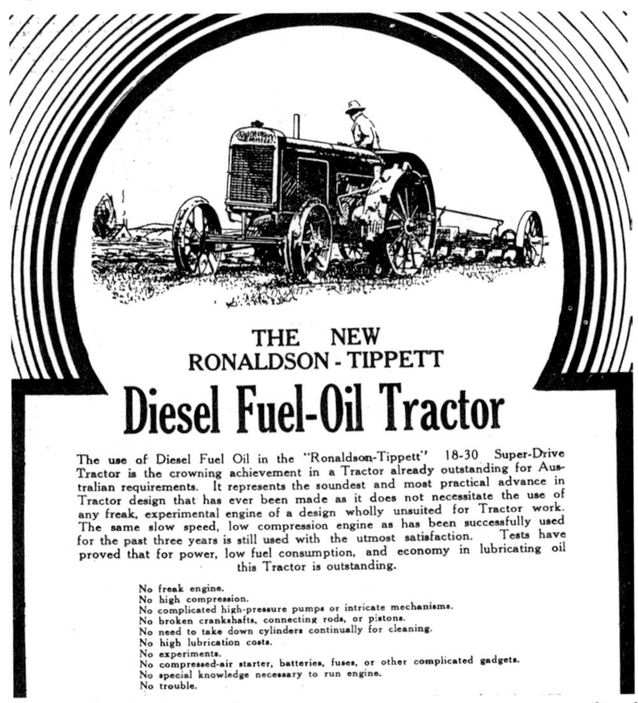

Not a diesel, but with serious imputations against true diesel engines. This advertisement appeared in Power Farming *in January 1928. RBT modified the manifold of their Wisconsin-licensed spark-ignited engine locally so the 18-30 tractor could be operated on cheaper 'diesel fuel' oil. RBT bulletins claimed that their 'patented diesel fuel oil vaporiser does not necessitate the use of any freak experimental engine. It remained to Australian engineering brains to successfully solve this problem.' Diesel fuel oil was about 2 cents per litre compared with power kerosene at twice the cost.*

*The producer-gas (charcoal suction gas) generator model of the 20-36 SuperDrive cost £595 in 1932. 'Turn waste timber and Mallee roots into power at a fraction of cost of other fuels' was the advertised claim. (*Power Farming, *1932)*

RBT SuperDrive advertising. (Power Farming, *October 1933)*

The 20-36 SuperDrive Tractor on rubber in 1937. (Power Farming Technical Annual, *1937)*

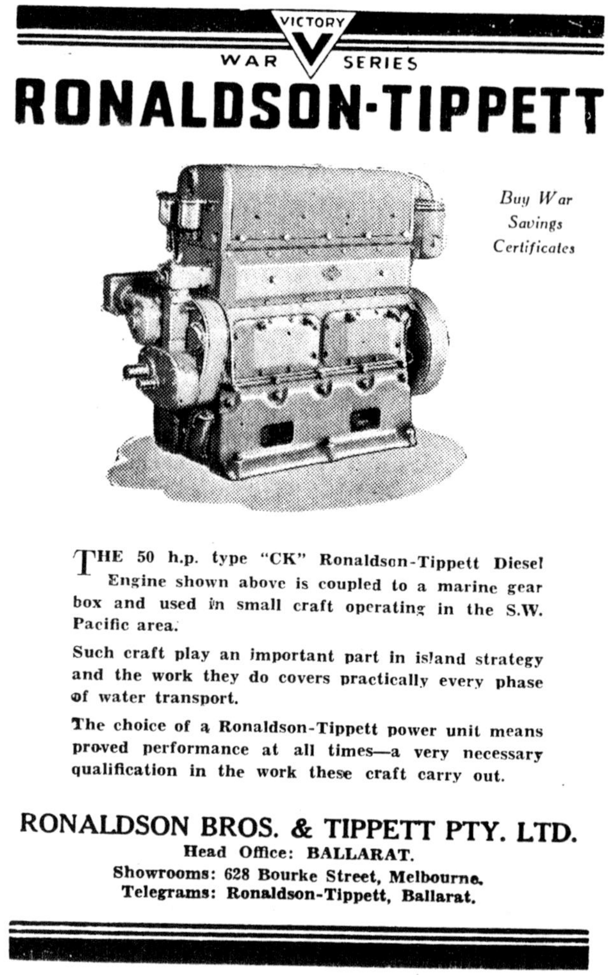

The Model CK 50 HP Ronaldson-Tippett engine for wartime use. (Power Farming, *1942)*

shed or even over weekends. Gil Tippett (nephew of Jack Tippett) tells how he had seen many a well-blackened shed roof. Despite this practice, diesel engines could still be operated more cheaply than the easier starting spark-ignition engines running on kerosene.

Many of the SuperDrive tractors were sold by the company on terms. This was before the existence of finance companies. When the Great Depression enveloped the country, the firm was greatly embarrassed by farmers unable to keep up payments on their tractors. The firm had to repossess many units and the financial strains of the period nearly broke them. Some of the repossessed tractors were rebuilt at the factory, renumbered and sold again, some as often as three times. Nevertheless, the company continued to advertise their tractors in *Power Farming*. The 20-36 model SuperDrive advertised and listed in the 1937 *Technical Annual* was fitted with rubber tyres. RBT producer-gas generators were also available through the 1930s.

Tractor production was discontinued in the late 1930s. According to Phil Wyndham, tractor buff in Western Australia, a total of about 340 RBT tractors was built. From there on the company concentrated on engines. They made single- and twin-cylinder vertical totally enclosed diesel engines for home-lighting plants, shipping and aircraft beacons. This was prior to the spread of the State Electricity Commission's power grid in Victoria.

During World War II, RBT manufactured four-cylinder diesels which were used in tandem or in sets of three twins to drive the three propellers of army supply barges for the New Guinea war campaign. The Ford Motor Company made the hulls in Geelong. There was also an air-cooled petrol engine developed for wartime tropical use. Before release this engine was tested inside the foundry core oven for its ability to withstand heat and dust. Munitions and gun castings made by the company also contributed to the war effort.

[I am indebted to Mr Gil Tippett, farmer of Relubbas via Horsham, for his contributions to this chapter. Gil Tippett is a nephew of the late Jack Tippett. Phil Wyndham of Western Australia also provided useful information.]

13 FRANK BOTTRILL'S DREADNAUGHT WHEELS

The fact is, there are no big jobs; only small machines.
— *R.G. Le Tourneau, in* Mover of Men and Mountains, *1972*

Arguably the physically largest tractor ever built in Australia was the brainchild of an Aussie battler, Frank Bottrill. The Adelaide-born blacksmith moved to Broken Hill in the 1890s, where part of his employment involved operating steam engines for the mines. He would often see Afghan camels transporting goods for outback stations and was distressed at the sight of the loads some of these camels carried. Some were laden down with burdens as large as four bales of wool (a load of more than half a ton).

Bottrill believed there was an opportunity for more efficient transport from the stations and more humane use of the camels by using steam power. He formed a family company to finance a steam engine enterprise and set out with his first engine from Adelaide, headed for Broken Hill. The tracks across the barren, treeless, sand-strewn country north of the River Murray were barely suitable for camels or bullock teams, let alone a heavy steam traction engine trailing a wagon. Inevitably, the outfit became inextricably bogged in the sand, right up to the boiler plates. In the absence of any trees to take the pull of a winch and cable, Bottrill's dream turned into a nightmare. He was forced to make the fateful decision to abandon his outfit in the desert north of Morgan. There it remains to this day, covered by drifting sand.

No doubt Bottrill thought about many alternatives to obtain trafficability across that arid region of Australia while he trudged disconsolately back to civilisation.

Frank Bottrill (1871–1953), blacksmith, engine-driver and tractor-builder.

Bottrill's Solution

The solution that eventually came to him was the 'Dreadnaught' wheel (also called 'Pedrail'): 'an improved road wheel for vehicles and travelling machines, especially useful for traction engines', so Bottrill claimed in his November 1906 patent. It consisted of flanged steel wheels equipped with a series of bearers which plopped down onto the ground for the wheel to run over. These bearers were fastened to the rim by steel cables. Bearers were alternated in two rows either side of the split rim. This was to ensure that the wheel rolled more smoothly on a rail as each adjacent bearer was picked up and carried around the rim. Bottrill's wheel was superior to a similar concept, the Boydell wheel, used between 1854 and 1870 in England on Burrell steam traction engines. Boydell's 'endless railway' laid down plates in just one row and, being intermittent, was noisy and more prone to wear, but at least it was the world's first commercially available tracklaying device.

This wheel system and many others that were developed overseas never survived as contenders against the crawler track, which was first made commercially available in the US in 1904 by Benjamin Holt. However, crawler traction engines were practically unavailable in Australia before World War I.

Frank Bottrill did not lose his vision of tackling the Broken Hill run. Between 1906 and 1910 he purchased first one, then a second McLaren steam traction engine which he equipped with his Dreadnaught wheels and contracted with others to carry out Mallee land-clearing and scrub-rolling work for the South Australian

One of the four steam traction engines used by Frank Bottrill and his brother Reuben for land-clearing contract work, 1906–10.

Bottrill steam traction engine pulling a 60-foot mallee scrub roller at Pinnaroo in 1910.

government. He pulled three rollers, of combined width 60 feet, and could clear up to 50 acres a day. The South Australian government, which had opened the area, charged the new settlers three shillings an acre for this work, which could be repaid when the land came into production. This style of opening up new land appealed to the Queensland government, so they followed suit with their own contractors.

Before long Bottrill began a business of equipping new oil-engined tractors, as well as steamers, with his wheels, for mining, log-hauling and rail track construction as well as his own land-clearing work. In England, John Fowler and Sons assessed 'Bottrails' (as they called them) on one of their steam traction engines. This was in response to an offer by the British War Office of a £1,000 prize for an all-terrain machine for military purposes. Hornsby & Sons developed a chain track and received the prize; Bottrill had no orders from England for the Bottrails.

Between jobs as a land-clearing contractor, Frank and Reuben Bottrill worked on installing Dreadnaught wheels on tractors purchased by mining interests as well as the Commonwealth and State governments. Shakedown trials were often carried out on the sand dunes at Fishermens Bend. There the Pedrail-equipped tractors demonstrated their ability to haul out bogged steel-wheeled tractors.

In 1915 Bottrill, encouraged with financial support from R.S. Falkiner, the grazier, embarked on his biggest venture. Big Lizzie was a 34-foot long, 45-ton tractor powered by a 60 HP Blackstone single-cylinder crude oil engine. The designer constructed the frame and assembled the tractor himself at a yard rented near the A.H. McDonald & Co. works in Melbourne. Gears and other machine components were supplied by the McDonald company.

*This tractor and trailer, built by Austral Otis for the Mount Gunson copper mines, was required to be fitted with Bottrill Pedrails; McDonald 45 HP engine powered (*Weekly Times*, 26 August 1911).*

Frank Bottrill's wife named this tractor Zena and that was the name used by the locals thereafter. The McDonald Imperial Model EC tractor with Pedrails was built for Federal Mines Ltd, Cloncurry, Queensland. It is shown loaded up for its first trip to the mine site with Frank Bottrill on 14 June 1913. (Jack Gates photo collection)

Bottrill's ambition was still his original desire to transport goods and services to and from the stations around Broken Hill. To this end, he constructed two wagons to pull behind Big Lizzie, each 30 feet long, with a capacity of 35 tons and each equipped with Dreadnaught wheels, as on the tractor.

The 'desert train' that left Melbourne in 1916 was a rig capable of hauling an 80-ton payload, and carried 17 tons of fuel oil. The family team reached Mildura in 21 months, after an epic journey. Where roads and bridges were

Big Lizzie arriving in Mildura with the Bottrill family and team in 1917. The 34-foot long tractor hauled two wagons, each of 30 feet. The wagons were also equipped with Dreadnaught wheels. Top speed of the whole outfit was barely one mph.

Big Lizzie hauled a record load of 75 tons of wheat from a Merbein farm in 1919. This was 899 bags — the whole of the farmer's crop. The front axles of each of the wagons were steerable.

inadequate, the Bottrills made their own. It took 200 feet and 40 turns on the wheel to swing the machine around a corner.

When Bottrill reached the Murray River it was flooded. What a setback! He had been without income for three years, so he had to find local work for the rig. Wheat carting gave him an opportunity to set a record for a single load of 75 tons of wheat when he brought in 899 bags to the Merbein railway station in 1919. When asked why he didn't round it off at 900 bags, Bottrill responded that 899 bags was the farmer's entire crop.

Scrub clearing became Bottrill's main preoccupation after that adventure and he developed a technique uniquely suited for Big Lizzie. It involved a team of up to 16 men, working with two 1-inch steel cables 100 feet long — one behind and one in front. The cables were used for tree felling and stump-pulling while moving forward and reversing. If repairs were needed these were speeded up en route by using the complete blacksmithy — forge, anvil and all — fitted out on the foredeck of Big Lizzie.

Some of the period photos of the machine at work reveal a large hook. This 'big hook', as the men on the team referred to it, was another Bottrill concept, developed especially to grub out ground-flush mallee stumps. The spring-steel hook, which weighed 64 kg, could be a lethal weapon with the tractor straining on the wire cable to overcome a stubborn stump. Bottrill was a cautious operator, however, who carefully drilled the crew, and no one was ever struck by that hook — there was no workman's compensation in those days. In spite of the arduous work and all his personal frustrations at not yet crossing the Murray to tackle the Broken Hill run with Big Lizzie, Frank Bottrill, a faithful Seventh-day Adventist, was also patient. He was never known to drink, swear or to lose his temper.

Big Lizzie's foredeck carried a blacksmithy, complete with forge and anvil. Note the 'big hook' at the rear, fastened to a steel cable and used for dislodging mallee roots at ground-level.

Financial obligations caused Bottrill to turn south again in 1926. He commenced land-clearing work at Glendenning, south of Horsham, on one of the properties of Ralph Sadlier Falkiner, the financier who had assisted him to build Big Lizzie 10 years earlier.

One of the wagons built by Frank Bottrill was used to move two portable steam engines in 1923. This setup capsized on Red Cliffs irrigation country when attempting to cross a ditch.

machine's travel speed and work rates. Crawler tractors with rippers and dozers were starting to make inroads: they could travel much faster, were more manoeuvrable and did not need as large a crew. The Dreadnaught wheel was destined to secure its slot in history but not in the mainstream of commerce. Nevertheless, it was finance and principles, and not really the mechanical difficulties, that prevented Frank Bottrill from going to Broken Hill with Big Lizzie.

[Big Lizzie has been partially restored (although not with the original engine) and can be seen at Red Cliffs, Victoria, in the Town Park. The author is grateful for first-hand information provided by Jack Gates of Mildura. Gates is a nephew of Bottrill and has built two scale working models of Big Lizzie. The assistance of Ron Maslin of Temora, who has written a book about the man and his machine, is also acknowledged.]

The arrangement was that Bottrill's indebtedness could be worked off by clearing land at Glendenning and he would be permitted to farm that land to generate further income. As it turned out, Falkiner was overseas for an extended period managing an extensive sugarcane plantation and working to perfect a cane harvester. The Glendenning station management had difficulty coping with a man who would shut down his tractor and not work on principle on Saturday, the seventh day. Disagreements over finances and conditions finally forced Bottrill to sell what he could of his interests and in 1928, a disappointed man, he left Big Lizzie behind.

Big Lizzie had a major flaw — she could travel barely faster than one mph. The big tractor was equipped with six bearers per wheel, whereas Bottrill's earlier designs had up to 12 bearers per wheel. At speeds above one mph, the larger bearers on Big Lizzie would lever out against their retainer cables with such force as to become unmanageable. This placed a major constraint on the

Specifications of Frank Bottrill's Big Lizzie

Engine: 60 HP at 275 rpm Blackstone single-cylinder. Hot-bulb ignition semi-diesel, crude oil fuelled, compressed air start. Water cooled, with Bottrill-designed exhaust-induced airflow 'radiator'. Single flywheel 7 foot diameter, 3 ton weight; all-up the engine weighed 8 tons.

Transmission: spur and planetary gear drives, four forward speeds, two reverse. Big Lizzie was arguably the biggest tractor built in Australia, but it certainly had the largest gear levers. The two levers were each 10 feet in length.

Dimensions: tractor length 34 feet; width 11 feet; height 18 feet, with crane jib adding another 3 feet 6 inches.

Drive wheels 6 feet diameter at the rim, 4 feet wide, each equipped with six bearers 7 feet long and retained on the rims by three cables. Front wheels 4 feet diameter, 2 feet 6 inches wide with six 4 foot 6 inch bearers. Steering radius 200 feet, requiring 80 turns of the wheel from lock to lock.

Tractor all-up weight 45 tons. Carrying capacity 10 tons plus 650 gallons of fuel.

14 H. V. MCKAY'S SUNSHINE TRACTORS

But things are easier in the Mallee now. Then it was all horses, and you'd be literally working 16 hours a day.
— *Wilbur Howcroft, in* Boots and All, *1984*

Hugh Victor McKay (1865–1926)

No story about any aspect of the mechanisation of Australian agriculture could ever be complete without a mention of Hugh Victor McKay. Arguably one of Australia's greatest industrialists, McKay started building farm machinery early in his working life. He designed and completed his epochal stripper-harvester at the age of 18 and the design was patented in 1885. This harvesting machine, while not the first of its kind, made a significant contribution to Australian broadacre farming technology.

The 27 HP Model A Sunshine tractor, which was built around 1916. It featured an unusual-looking radiator style and was initially intended for use with mounted harvester, as shown. The tractor alone cost £500.

From humble beginnings, this resolute young man developed a business of manufacturing harvesters, starting in Ballarat. Over the period between 1904 and 1907 McKay moved shop to establish the Sunshine Harvester Works at Braybrook Junction, later named Sunshine (now a western suburb of Melbourne). During those years, while he was striving to establish the extensive factory complex, he also took on a massive publicity effort to fight 'Yankee pirates', as he called them. It was the International Harvester Trust that was the special focus of McKay's actions because he perceived them as a significant threat to Australian free enterprise — and McKay was the most vocal industrialist in his day. The Harvester Trust had all but monopolised the North American farm equipment industry at the time and was expanding its network offshore (see chapter 19).

The 15 HP Model O Sunshine tractor was finished stylishly for the period — 'along motor-car lines' — around 1917.

McKay Tractors and Self-propelled Machines

H.V. McKay manufactured a complete range of farm machinery at Sunshine. He experimented with a number of self-propelled harvesters, from around 1917 until 1923, when the first of the Sun Auto Headers was marketed.

Actually, tractors seem to have been a spinoff from these self-propelled harvester developments. A 1916 patent to H.V. McKay discloses two tractor designs, Models A and O. Both were tractor/harvesters, in which the tractor could be disengaged and used for general purposes. Both tractor designs were subsequently produced in limited numbers.

The 15 HP Model O Sunshine tractor had the traditional north–south engine layout, with a four-cylinder US-made Continental engine. The tractor was a clean-looking design, 'constructed on motor-car lines, except that the rear wheels are driven by pinion and segment' (*Land, Transport and Sea Trade*, October 1919).

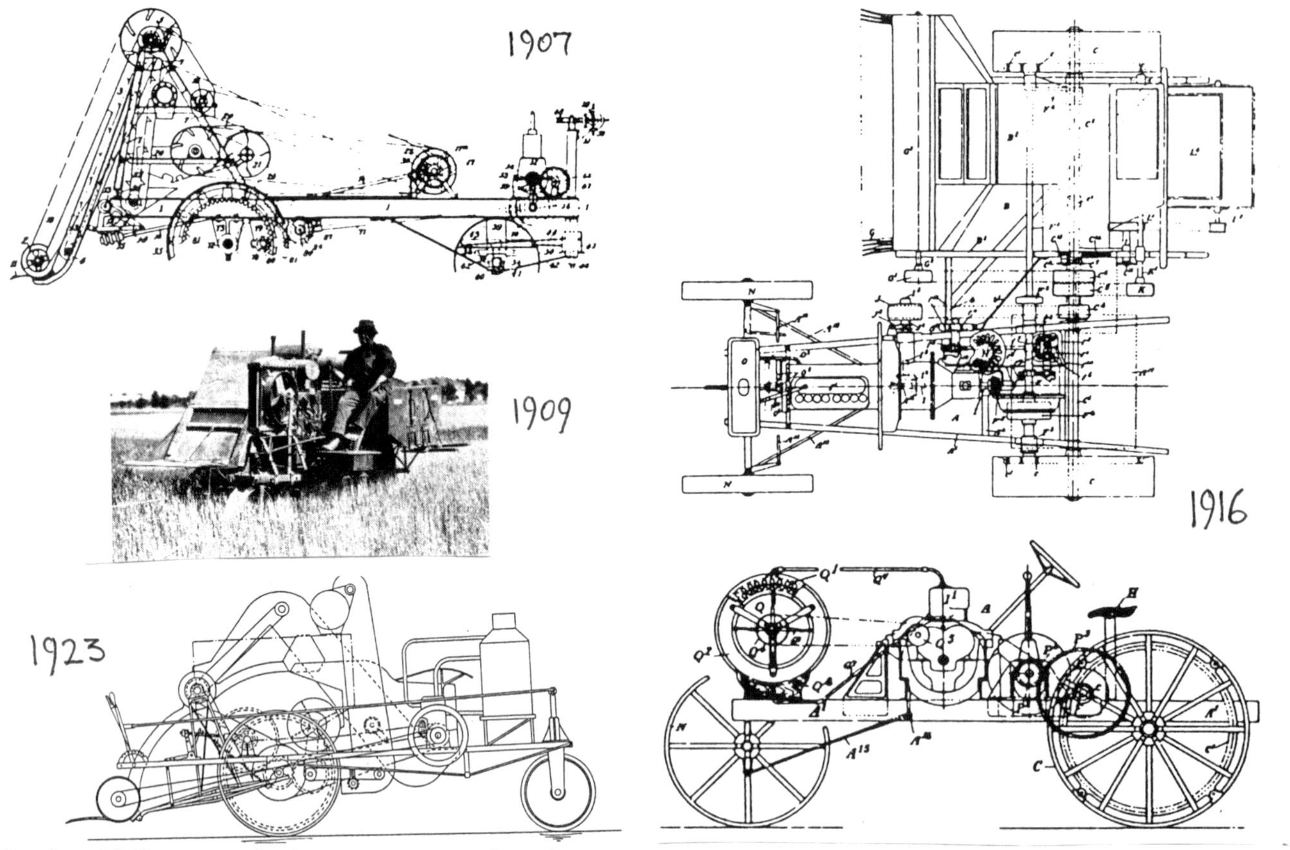

Sunshine McKay tractor development. Drawing above shows McKay's 1907 Australian patent Specification (No. 9,759) for a grain harvester attachment to a 'road tractor or traction engine'. The design was laid out so that the bagged grain could be carried to a convenient dumpsite, then the harvesting attachment removed so that the machine could be used as a road tractor for haulage of the bagged grain. There was also a self-propelled stripper-harvester with the engine at the front (1909). These machines did not go into production. A 1923 patent (No. 13,492/23) followed similar lines, but McKay's 1916 patent (below) was quite different in approach and was built around the Sunshine Model A or Model O tractors.

In 1984 Australia Post issued this commemorative cover, 100 years after H.V. McKay demonstrated his first stripper-harvester.

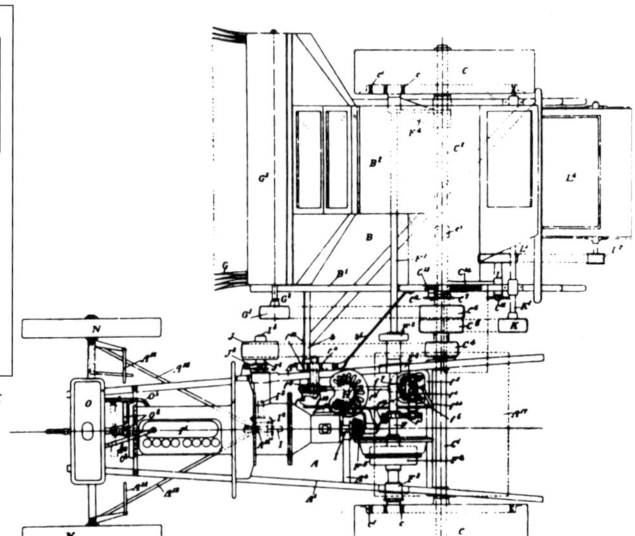

The Model A Tractor

In 1918 McKay entered one of each of his tractors, the Model O and the 27 HP Model A, in the Motor Tractor Trials arranged by the Royal Agricultural Society of Victoria on the State Research Farm, Werribee. At these trials it was recorded that the Model A ploughed 3.63 acres in 4.45 hours. Fuel consumption was 1.73 gallons of petrol and 6.0 gallons of kerosene. Some 15.6 gallons of water were required — twice the volume of fuel. Water was needed on most kerosene engines of the day as an anti-knock under load.

Special emphasis was placed on keeping weight down with the Model A. One advertisement labelled it as the Sunshine Light Tractor. All-up weight was just 1.8 tons. This time the US-built Continental Motor Corporation engine (a four-cylinder in-line four-stroke, with a bore size of $4\frac{1}{8}$ inches, stroke 5½ inches and 294 cubic inches displacement) was mounted transversely on the Newcastle steel chassis, on cast iron mountings made at the Sunshine works. The tractor carried a fuel tank with two compartments — one for petrol to start and warm up the engine, the other for running on kerosene — and a water tank. Two carburettors were fitted. McKay claimed the engine would also run on alcohol. Ignition was by high tension magneto. A water pump circulated water through the annular radiator headers to the cage of interconnected pipes enclosing a fan similar to those used in the Sunshine threshers and winnowers.

The differential had straight-cut, hardened gears running in an oil-tight casing while the final drive to the 4-foot diameter steel wheels was by pinion and ring gear (as used on the stripper harvester). Steering was of the automotive type. McKay sought to use as many common parts as he could from his other product lines parts inventory. The retail price of the tractor was £500. By comparison, the Model O retailed at £350 in 1918, the same price as the popular 16 HP IH Mogul.

Both Sunshine tractor models had a three forward speed/single reverse gearbox and drove the wheels by externally mounted pinions and ring gears, as on the harvesters.

Very little is on record about H.V. McKay's tractor ventures—either by reportage or advertising. This absence of coverage in the rural press is strange because McKay was never reluctant to publicise his other product lines. McKay exhibited tractors at the 1919 Royal Melbourne Show. Unfortunately no records remain to show exactly how many Sunshine tractors were made. A 1917 report indicated that a dozen specimens were in various stages of construction and that the firm intended, if demand was sufficient 'to produce the machines by scores and hundreds'. A 1949 company publication suggests that 'quite a number were sold, including one which was exported to Egypt'. For reasons unknown, production stopped some years before the McKay's death in 1926. Perhaps the lack of promotion was a consequence of the overwhelmingly low price of the rival Fordson.

The Museum of Victoria was fortunate to be able to purchase a Sunshine Model A in 1980, which it has fully restored. As for the Model O tractor, Phil Wyndham of Western Australia believes that there is one surviving in Tasmania.

Restoring the Model A

The Museum of Victoria's restoration work on the Model A involved many challenges. The most difficult was to locate information on the unusual radiator. The radiator was missing from the tractor, which was otherwise substantially complete when purchased, although many components (including the engine) had been disassembled for repair work. Nothing could be located on the radiator in the surviving company records. Eventually, sufficient details of its construction were found in patent records, but these did not specify the materials used. After widespread consultation, a decision was made to use copper tubes for the reconstruction. Subsequently a description of the tractor came to light from sources outside the company, revealing that McKay had used 'ordinary galvanised iron pipes'.

Determining the proper colours of the paint finish was another challenge. Although clear traces of yellow paint were evident on all wheels, no traces were found on any other components. What appeared to be gray paint was later identified as iron oxides. The earliest oral evidence on colours was obtained from Mr Les Shea, of Connor-Shea fame, who started at the Sunshine works in 1924. He remembered a Sunshine A tractor in store at the works and recalled its colours as 'predominantly green with yellow wheels'. He suggested that it would have been painted with the same green, yellow and red used for the Sunshine harvesters and implements. Colours matched to the museum's Sunshine harvester were therefore used as the basis for the colour scheme on the restored tractor. The original magneto was missing from the engine and was replaced with one of a similar style. Because the governor was also missing, direct throttle control has been fitted until a suitable

Left and right views of the Sunshine A tractor.

replacement can be located. In restoring the engine, additional oil feeds were installed to the centre main bearing, which had seized. It is not clear whether the museum's tractor originally had a canopy. Because of the protection it offers, one was fitted, based on the example of the Sunshine A seen in photos of the 1918 tractor tests. This unique tractor is displayed at rallies.

Without fanfare, Dave Berthelson of Gayndah, Queensland, has located and restored a Sunshine model A tractor, which he exhibits and rallies.

Dave Berthelson's Sunshine Model A, Gayndah, Queensland. (Photo courtesy Kevin Burnett, Gayndah Shire Council)

[I am greatly indebted to Martin Hallett, Museum of Victoria, for the illustrations and for much of the material in this segment.]

The rural press gave little coverage to the Sunshine tractors, but this advertisement did appear in Land and Transport *in 1919. McKay had never previously been reluctant to publicise his equipment lines. Perhaps it was the low-priced Fordsons that caused McKay to hold back on the Sunshine tractors.*

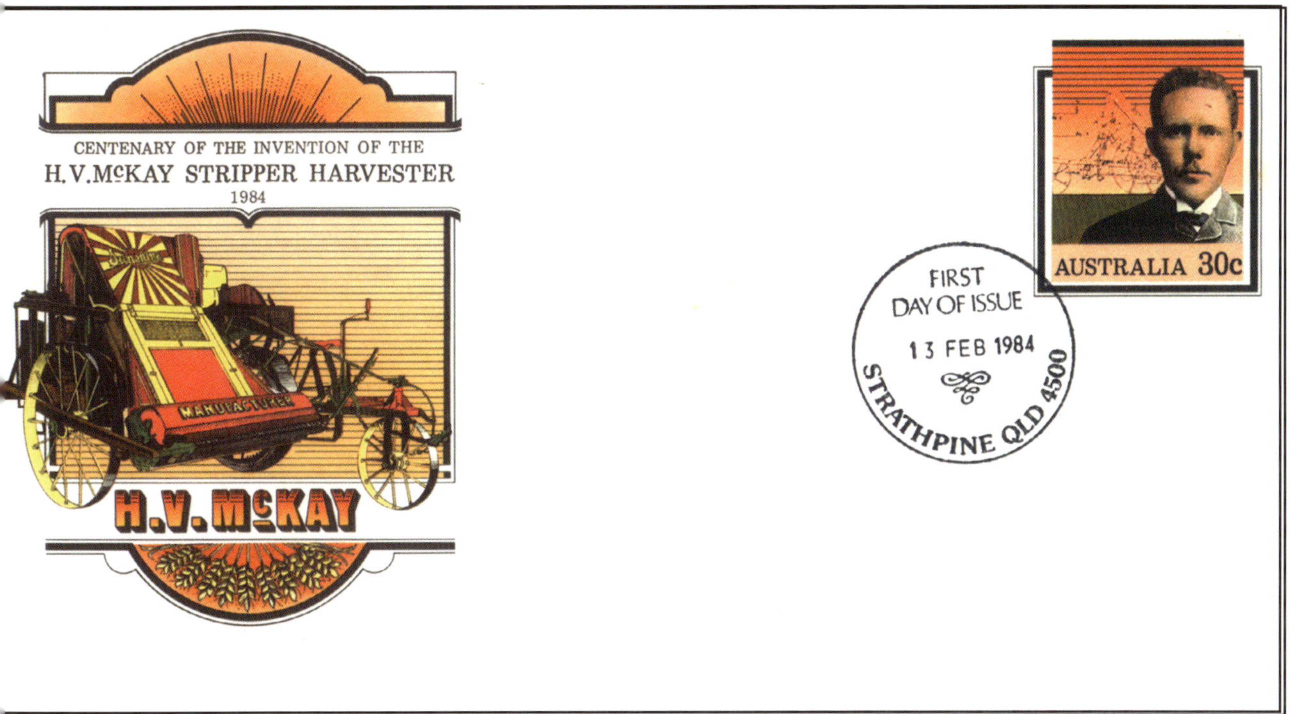

cKay Stripper Harvester, 1884 and Hugh Victor McKay, as seen on a centenary celebration first day cover issued by Australia Post in 1984.
acKay went on to develop a multi-national farm business and added some tractors to his line.

1930 Vickers-Aussie tractor in the Whiteman Museum collection, WA. (Photo courtesy The Old Machinery Magazine, *December 1997)*

Tail-end Charlie, a Chamberlain Champion specially modifi for high speed road work, is in working condition in the Whiteman Museum in WA.

Tail-end Charlie could reach 68 mph. This tractor followed several round-Australia rallies, having to cover a distance of 11,140 miles to complete the circuit in 19 days in the 1957 Mobilgas Trial.

Author's photo of one of the Hart-Parr Australian Special Model 12-24Es. This well-restored tractor was at Queensland's Jondaryan Woolshed Museum.

When you visit the Jondaryan Woolshed be sure to inspect one of the most unusual tractors ever to be seen. Irvine's 37-ton Road-Burner was a mobile kiln that was used in the late 1920s to bake clay roads to make them trafficable in wet weather in outback Queensland. A team of 23 accompanied the machine, which worked 24 hours a day on the job and burned whatever fuel was available. It travelled at 220 hours per mile! (Author photo)

KL Bulldog 40 HP crude oil tractor. The Bulldog was restored by Bert Den Hartog of Camden, NSW, who provided the photo.

KL Bulldog driven by Steve Muscat of Riverstone, NSW, helping out a stalled Field Marshall during TOMM's Inaugural 1998 Wingham Classic Tractor Trek. (Courtesy The Old Machinery Magazine*).*

15 CLIFF HOWARD, THE TRACTOR BUILDER

[Howard's] DH22 tractor, designed in 1927 to work with rotary hoes, initiated the first large-scale production of tractors in Australia.

— Australian Dictionary of Biography, *Vol. 9, 1983*

Cliff Howard is world-renowned for his rotary hoeing machine, the Rotavator, which was his brainchild. Less well known is the fact that he was the pioneer of production-line manufacture of tractors in Australia. There is a reasonable association: tractors are essential for rotary hoes. Cliff Howard knew this, and when there was a faltering in the supply of Fordsons prior to 1928 he decided the time had come to really get serious about the business of serial production of his own tractors to carry Rotavators.

But to go back a few years. Cliff Howard was just 16 when he built his first powered rotary cultivator. This took place on his father's farm at Gilgandra in 1912. After many experiments, he felt that he had a design worth marketing, a rotary cultivator with L-shaped blades:

'I put my motorcycle engine on our crude prototype to demonstrate its ability but there was little or no interest in a small machine as market gardening was nearly all done by Chinamen who had no money or ambition to use anything but a hand hoe.

'The only people I was acquainted with who showed any interest in my new method of tillage were the big wheat farmers and obviously small outfits were of no use to them. They were used to working with horse teams of 8–12 and anything less in working capacity held no interest, so we set ourselves the task of designing a machine that would cultivate the land fast enough to match or better these big horse teams, having a seed drill incorporated so that seed sowing could be carried out at the same time and the chassis of the machine so constructed that the rotary hoe could easily be detached and a combined reaper and thrasher mounted in its place — a complete wheat farmer's outfit.

'However, the work I was engaged in with the firm where we were employed [McCleary's of Moss Vale] kept me very busy and as it was chiefly away from the home town our progress with preparation for our ambitions was slow.

'By the middle of 1914 we [Cliff with brother Albert and a fellow apprentice] had the first part of our scheme just about ready but, about this time the 1914–18 war broke out and our firm went straight into war work. We decided to postpone our venture and volunteered for the air force. Whilst awaiting call-up I had a motorcycle accident which resulted in my being turned down so I later went to England and worked on munitions and aeroplane engines until the end of the war' (Cliff Howard's reminiscences with J.P. Richards).

Arthur Clifford Howard (1893–1971).

In 1919, when his contract had run its course, Howard returned from England. He had not found any backers over there for his ideas, in spite of having contacted many British agricultural firms. Early in 1921 he formed a small syndicate in New South Wales to finance a rotary hoe venture. The prime mover for the wheat cultivating machine was a Halford lorry-engine of 40 HP. Encouraged by the way this machine performed, Cliff Howard imported a US-built 60 HP Buda engine to power a 15-foot wide rotary hoe, made of five sets of rotary hoe gangs, each three feet wide.

Howard's First Tractors

Howard had originally planned a universal machine. It was to cultivate and sow seed, with the truck-type chassis layout designed so that the rotary hoe rig could be removed and a reaper-thresher mounted in its place. The 60 HP machine plowed 3½ acres in 1922 and, as a result of orders coming in from demonstrations on New South Wales wheat farms, plans were made to

One of the early Howard tractors, c. 1922. The five rotary hoe gangs could cultivate a 15-foot swath. The tractor was powered by a 60 HP Buda engine. There were six travel speeds and three rotor speeds.

been located and is restored at Moss Vale.

In January 1923 Howard registered the business name Austral Auto Cultivators Ltd, and set about selling shares in the enterprise. His shareholders consisted of farmers, businessmen and friends from far and near. Of particular note is entry number 82 in the company's original share registry, which shows that the Hon. W.M. Hughes, barrister-at-law, took 1000 shares worth £1 each in the company. This was none other than Billy Hughes, who later became Prime Minister of Australia, pushed for conscription through the years of the Great War and led the country for several years after.

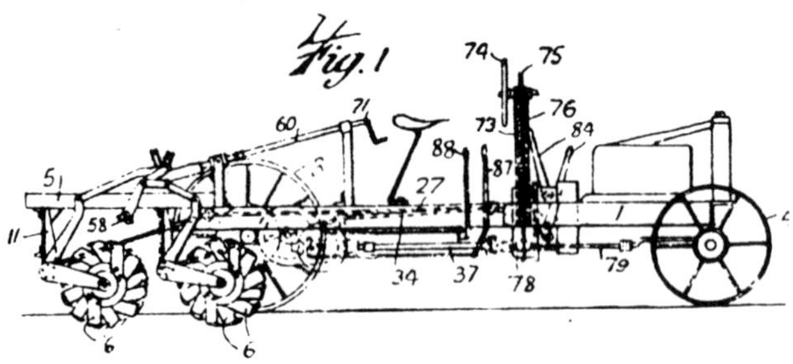

Early photo of one of the six original tractors, with its five gangs of Rotavators. The Howard brothers, Cliff and Albert, are on board.

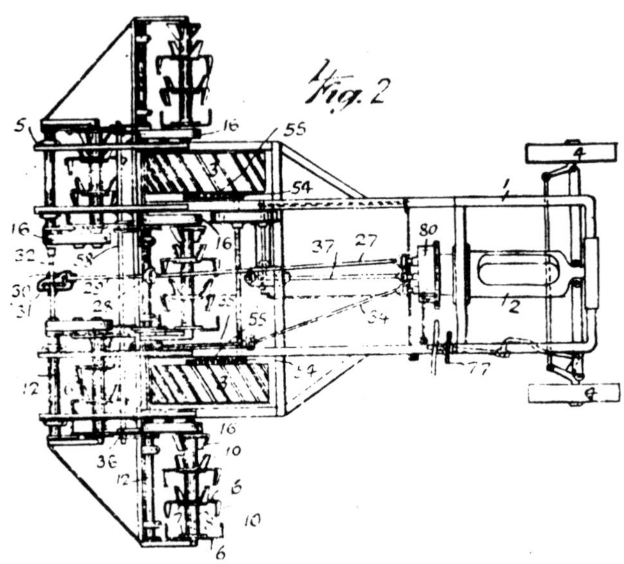

Howard's 1920 Patent (No. 18,137/20) shows the three-way power splitter from the engine to drive the rear axle and cultivator gangs, each drive having its own clutch.

build six of these tractors. The 1920 Australian patent awarded to A.C. Howard on this tractor concept reveals a splitter transmission dropbox mounted on the engine bellhousing to give three power outlets to drive the tractor rear axle and the outrigger rotary hoes. Each outlet had its own clutch. One of these machines has

With this share capital, Howard acquired the old Moss Vale factory where he had once been apprenticed. Selling each tractor and cultivator rig was mandatory in order to pay for the construction of the next one, and collecting dues was no mean task:

'On one occasion my man and I arrived at a farm to deliver a machine after having driven it 50 miles from the nearby railway station, chiefly over unmade roads in rainy weather with many creeks in flood. We got bogged so many times it took two days to do the journey and an arrival found the farmer in a very bad mood as the machine was a month late. He had missed all the good ploughing weather. He thought it was too late to do any good with his crop that season and wanted to get out of taking the machine, after seeing a five-acre patch tilled in 1½ hours he agreed to take delivery and pay for the unit if my man and I stayed on the job and ploughed all the land he had ready to crop — which amounted to about 1000 acres! It was most inconvenient as I had other urgent jobs to attend to, but we had to have that money to keep going, so we started in right away and kept the machine going continuously, stopping only for oiling and refreshments for 15 days till the job was done.

'Another time a noted doctor expressed interest in our company and arranged to see a demonstration. He motored 100 miles to where the machine was working and was greatly impressed. I had made out an application for 1000 shares ready for him to sign, then he thought his wife might also like to take some shares so he went to the car and spoke to her about it. She got out of the car to take a closer look at the work but her Pekingese lap dog, which she was carrying, seeing the blades revolving then disappearing into the ground, jumped out of her arms and made a grab at a blade. He was, however, not quick enough and the next blade caught him, cutting his head clean off. The poor lady was so distressed she collapsed with grief and nervous prostration and had to be carried back to the car and taken home at once. I never saw the doctor or his wife again and with sad hearts we buried the Pekingese' (Cliff Howard's reminiscences with J.P. Richards).

Howard's early experience with motorcycles proved invaluable when he started to make small air-cooled engines for walk-behind rotary hoes in 1930. Smallest of the Howard product line in the latter part of the 1920s was the Junior with Howard 5 HP four-stroke engine. This single-axle tractor had three forward speeds. Initially engines were imported from England, but problems with reliability and spares motivated Howard's engine ventures.

The six tractors with their 15-foot cultivators were sold by the end of 1923, although the seed drill and harvester attachments were never built, due to lack of finance.

Fordson Fling

Cash-flow problems forced Howard to look to alternatives. The Fordson tractor was rapidly rising in popularity at the time, and it looked attractive for rotary hoe work, although a low-speed or creeper gear was unavailable. With astute and extensive marketing of his attachments for Fordsons, especially to orchardists and canegrowers, Howard's business grew to the stage that in 1927 it required a move to a bigger factory, nearer to raw material suppliers. Northmead near Parramatta was chosen and a new factory was built there by the energetic inventor.

The Fordson tractor application proved most lucrative for rotary hoes, but Howard diversified into small single-axle or walk-behind rotary hoe tractors. These were the Junior (5 HP), the Eight and the Twelve, powered by air-cooled engines initially imported from England. There was also the Rolling Dolly hand-operated washing machine (only 40 built). In 1928 Howard Cultivators Ltd was licensed in England to handle export orders outside Australia and New Zealand to match the growing interest and demand that was developing offshore. In that same year the first Howard cane harvester was built, along with a helical mouldboard plough, and a three-wheeled tractor. The three-wheeled tractor, known as the Auto Rotary Hoe 16, was powered by a 16 HP Morris engine and carried a mid-mounted three-foot Rotavator. Later in 1928 came a setback for the company: the Fordson tractor was suddenly withdrawn from the market.

During the next two years, while Fordsons were unavailable, there was practically no market for Fordson attachments. Howard had introduced creeper gears and power take-offs as well as rotary hoes for the Fordson. So the push was on to market the single-axle tractors. A variety of engines was tested, each having its shortcomings for rotary hoe work, and there were also problems obtaining spare parts. Cliff Howard decided to design and build his own engines. Air-cooled engines were produced first for the Junior and Eight.

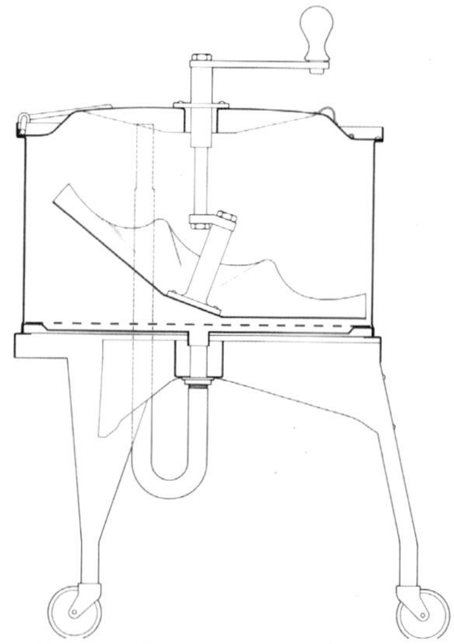

Rolling Dolly — Howard's washing machine, c. 1928. Only 40 were built.

For the Twelve, the Sixteen, and a new four-wheel tractor, the DH22, a series of water-cooled engines was developed. The first models of the DH22 (the prefix stood for Daisy Howard, the designer's wife) were fitted with Morris Commercial engines of 22 HP. To establish a foundry for engines, more land was acquired at Northmead, and the company name changed to Howard Auto Cultivators Ltd as part of a push to raise more capital.

This was the Howard exhibit at the Royal Agricultural Show, Sydney, in 1932.

The DH22 tractor, launched around 1930, became the company's tractor line-leader and was manufactured for the next 30 years.

More uses for Howard engines were found: engine power packs were advertised for fitting to harvesters to increase capacity, and as universal power units for other work around the farm. Ever fascinated by engines, Howard even attempted a Ricardo-type diesel in 1932 but was distracted by business hassles from persisting with this development. During the 1930s Howard devoted a considerable effort to capture the sugarcane machinery market with his hoes, tractors,

The DH22 tractor became Howard's tractor line-leader for 30 years after it was launched around 1930. The earliest version was powered by a Morris 22 HP engine from England. Later the company produced their own water-cooled engine for the DH22. The Howards often conducted tests at Hawkesbury Agricultural College, where this photo was taken before World War II.

At the Buenos Aires farm machinery show c. 1929, Howard representatives displayed the Sugarcane Special three-wheeler Rotavator, the Sixteen, and a Fordson-equipped helical plough, or 'rotary mouldboard'. Howard patented this helical plough in 1928 (Australian patent No. 11,648/28).

Howard's locally built four-cylinder spark-ignited engine of 22 HP, for the DH22 tractor. It was reportedly the first overhead-valve engine manufactured in Australia. All basic components were manufactured at Northmead, except for the imported magneto and ball bearings.

This three-wheeled Howard tractor with 16 HP Morris engine was built primarily to carry the mid-mounted Rotavator. It was manufactured between 1926 and 1930. The patent (15,351/28) was issued in 1928. There were actually three models: orchard, sugarcane and field.

Howard Auto Rotary Hoe 22 with Howard engine and rubber tyres, c. 1937. The tractor had 10 forward speeds, starting at 0.9 mph creeper speed for rotary hoeing. Hydraulic hoe lift was featured in 1935.

D16 Morris-engined canefield model with side shields was used to drive a 3.1-foot Rotavator, the Auto Rotary Hoe 16.

Special DH22, modified at the factory to carry the Howard producer-gas generator, around 1937.

engines and a self-propelled cane harvester.

By 1934 the Northmead works boasted a staff of 140 and, despite the growth of the business, A.C. Howard still could be found at the workbench on new inventions. During the lean period of the 1930s farmers were attracted by any production method that might save money, and producer gas was a much cheaper fuel than liquid fuels. Howard tested producer-gas generators and built a 40 HP tractor prototype with an engine specially modified for producer gas, but with the capability to run on kerosene if required. It was first tested on 600 acres of wheatland at Gilgandra, then sold to a Bundaberg cane farmer.

In 1936 a number of DH22s with slightly modified engines were fitted with producer-gas generators to find out how they would perform in the hands of farmers. Most purchasers started off well but they soon tired of the constant attention the gas producers required. Many decided to convert back to kerosene even though the fuel cost per acre was over four times that of producer gas. After that, no more tractors were sold with gas producers but experiments continued to improve on the design.

In 1937 the DH22 was offered with pneumatic tyres. In that same year (the year *Power Farming's Technical Annual* was launched) Howard's offered a four-wheel-drive industrial tractor, the FWDH22. It was designed with two steering wheels so it could be operated in either direction if the operator switched position. There were four speeds forward and four reverse gears. For shunting

The D16 four-wheeled tractor replaced the three-wheeler, c. 1929, but still used the 16 HP Morris engine.

rail wagons, this arrangement saved turning the whole tractor around. There are hints of the French Latil tractor here.

Troubles developed in England in 1934, when the British subsidiary decided that they would take out patents in their own name. Cliff Howard went over in October 1936 to sort matters out, then again in 1937, but during his absence while working out the legal issues, management troubles caused a rift in the Australian enterprise. As it happened the British episode turned in Howard's favour and he formed a new company in England, at the same time divesting himself of any involvement with the Australian management. Rotary Hoe Cultivators Ltd, England, was registered in 1938. For Cliff, it was pioneering all over again. He pitched in to level the site for the works at Horndon, Essex, and even personally laid bricks for the new factory. Not long after, however, a financial downturn plagued the British company and it was barely salvaged by the outbreak of World War II, when the works was devoted to producing munitions and military equipment.

A four-wheel-drive model, the FWDH22, using the same 22 HP power unit as the DH22, was offered in 1937.

Meanwhile the Australian firm's attention was also shifted to the provision of war materiel and the Northmead factory made ammunitions and even a revolver, known as the H.A.C. revolver. It was the only Australian-made handgun of the war. Back in Britain towards the end of the war, the government directed the Howard factory to resume agricultural production to help with the war effort.

Postwar Recovery

Immediately after the war, Cliff Howard worked aggressively at establishing a US market, selling products from his British company, Rotary Hoes Limited. In 1959 Howard returned to Australia to buy back the Australian company. A lightweight four-wheel tractor, the Model 2000, with 12 HP L-Series engine, was launched in 1962.

A Howard lineup around 1937 — four sizes of rotary hoes and three Howard tractors. The 5 HP Junior, 9 HP, the DH22, and the Caterpillar 75 with the 8.5-foot wide Howard rotary hoe which could cut to a depth of two feet in North Queensland canefields.

One of the 20 FWDH22s purchased by the NSW State Rail Authority.

machines apart from the Rotavator. Especially notable were a giant square hay baler (one-tonne bales) and the slant-leg Paraplow. So the Howard name survives, although under different organisational structures on each continent.

[The assistance of J.P. (Pen) Richards, Killara, with textual notes and the many photos he lent for this segment is gratefully acknowledged. Material was also provided by Peter Milne, of Howard Australia (1985) Pty Ltd.]

Platypus Tractors

There was also an interlude in England with crawlers when the Platypus Tractor Co. was established to make Platypus crawler tractors, between around 1950 and 1958, at Basildon, Essex. The most popular model in the series was the Platypus 30 with the Perkins P4 31 HP diesel engine. The tractor had six forward speeds and two reverse, and maximum drawbar pull was three tons. A special version, the Platypus Bogmaster, with extended track, exerted a ground pressure of less than 1.3 psi (a man's footprint exerts 5–7 psi) but only a few had been built by the time the British Howard group pulled out of the crawler market in 1958.

Cliff Howard was in England as managing director in 1970 when he was awarded the CBE for 'services rendered to agriculture' from the Queen, but not many months afterwards he died, in January 1971. Howard had earned a worldwide reputation. He had founded many production facilities and established an immovable place for the rotavator principle in world agriculture.

The Howard Twelve tractor and hoe, 1952.

The Howard Platypus crawler tractor.

Company in Trouble, but Name Survives

The Australian company, Howard Rotavator Pty Ltd, suffered some severe reverses with the downturn of the farm equipment industry in the 1980s. In July 1985 Howard Rotavator Pty Ltd at Northmead was wound up, after having produced more than 100,000 Rotavators. Local production of tractors ended. New management and a move to Seven Hills saw another company, Howard Australia (1985) Pty Ltd emerge. This company became the Australian outlet to handle equipment manufactured overseas.

Even after the passing of its founder, the famous Howard name appeared on a range of important

The Howard DH226 was offered in 1952, with a US-built LeRoi engine of 35 HP which enabled the tractor to work a 4 foot 6 inch hoe. The LeRoi engine was also made available in the FWD tractor.

In 1970 Cliff Howard was awarded the CBE. Cliff and Daisy Howard after the Award Ceremony.

Cliff Howard in England. He drove this restored DH22 at a 25-year company commemorative event in 1962.

The Howard 2000 compact tractor was first offered in 1962. The 12 HP ride-on garden tractor was a popular machine and had a range of locally built implements to match.

16 THE VICKERS-AUSSIE AND THE HART-PARR AUSTRALIAN SPECIAL

There were two tractor models that were Australian in name only. They were built overseas. They did share one other trait, apart from using the national epithet. They were both doomed by the Great Depression.

The Vickers-Aussie

When it first appeared in 1925, the Vickers-Aussie was a fairly basic design of wheeled tractor, made by Vickers Limited of Sheffield, England, specifically for the growing market in Australia. A 30 HP spark-ignited engine, petrol/kerosene fuelled, provided the power to the rear axle through a three-speed gearbox of the unit-frame tractor. It bore a striking similarity to the US-built International 15-30. One authority suggested it was made under licence. The International 15-30 was superseded by the McCormick-Deering 15-30 in 1924 (Gray, 1975).

Both the McCormick-Deering 15-30 and the first Vickers-Aussie had four-stroke four-cylinder engines, 4½ x 6 inch (bore and stroke) 30 HP at 1000 rpm and employed the 'moist air carburettor which comes into operation when the changeover to paraffin (kerosene) is effected' (after start-up on petrol). Water to prevent knocking under load was supplied from a separate tank. The main fuel tank carried 18 gallons of paraffin, the auxiliary tanks four gallons of petrol and of water.

The rear driving wheels were a distinctive feature of the Vickers-Aussie. They followed an Australian design principle, patented in 1925 by Messrs Kendall and James. Each drive wheel of the Kendall-James patent consisted of three sections, and between each pair of sections was placed a fixed 'comb' to flush out any soil and mud which might otherwise stick to the wheel rims and grips. The manufacturer claimed in advertising that: 'The Kendall-James wheel will not slip when the going is most difficult [and] will give off 100 per cent efficiencywhen the conditions are ideal'. A British journal waxed more eloquent: 'with its special wheels, the Vickers could work easily where an ordinary tractor would be bogged down and where a man on foot would sink into the ground "to the top of his boots". It opened up a new and brighter era for power-farming' (*Implement and Machinery Review*, 1925).

The Vickers-Aussie, 'British-built throughout', was last sold around Australia in 1930. At that time the distributors, British-Australasian Tractor Co. Ltd in Melbourne, were offering the Mark IV version, with 23/40 rating (drawbar/belt).

Hart-Parr Australian Special

This tractor appeared only briefly and there were probably two models marketed in Australia. These were the Iowa-built Hart-Parr tractors — the 12-24 Model E and the 18-36 Model H. They were the last to bear the famous Hart-Parr logo before that company was taken over in 1929 by the Oliver Corporation (originally incorporated as Oliver Farm Equipment). Hart-Parr were popularly credited with establishing the word 'tractor' in place of the expression 'gasoline traction engine' that was in use up to 1906 (see chapter 1 and Gray, 1975).

The Hart-Parr Australian Specials were manufactured in Charles City, Iowa. In 1929, 1482 units of the Model 12-24E were produced, bearing serial numbers 38,119

The 1925 Vickers-Aussie tractor, British-built, but 'manufactured for Australian conditions'. Right: the patented Kendall-James 'self-cleaning, all-season, non-slip wheel'.

A 1928 Vickers-Aussie tractor, as restored and displayed at the 1985 Henty Field Days. It was painted orange. There is more than a passing similarity between this tractor and the McCormick-Deering 15-30.

McCormick-Deering 10-20 — a smaller version of the 15-30, on which the first Vickers-Aussies were modelled.

to 39,601 (White Farm Equipment Co., Charles City, Iowa, personal communication, 1986).

There were some minor modifications to the 12-24 tractor to suit Australian conditions, including the distinguished radiator casting, which really set it apart. The Model 18-36H was also advertised but how many were Australian Specials is unknown (4,920 units were built at Charles City in 1929). When Oliver took over the Hart-Parr organisation during the Great Depression these models were among the casualties, being unceremoniously dropped from the Oliver tractor lineup.

[There was also a New Zealand Special built for sale in that country from the same factory in Charles City, Iowa. Information on New Zealand Special from Doug Goddard of Tamworth.]

The Mark IV Vickers-Aussie was a more powerful, 23-40 HP model. Marketing of this tractor did not survive the Depression. Mark IV version advertisement in Power Farming, *25 November 1930.*

Left: Hart-Parr Australian Special advertisement, in Power Farming, *25 September 1930.*

17 THE KL BULLDOG

I think it does a woman good to sit on a tractor and do those jobs for a while to realise that when a man comes in he's really physically tired and it doesn't hurt to wait on him a bit. And if some women heard me say that, they'd raise their arms in horror.

— *Myra Robilliard, in* Boots and All, *1984*

The story of Australia's Bulldog tractor actually starts in Germany in 1921 with the firm of Heinrich Lanz AG, Mannheim. The Lanz business itself began in 1859 with threshing units and steam machines. The first Lanz Bulldog was designed by Dr Fritz Huber and it set the pattern of a simple tractor built around an engine of single horizontal-cylinder configuration, using the two-stroke principle with hot-bulb ignition. It was a semi-diesel engine that capable of burning a range of fuels that were cheaper than kerosene or petrol.

In Melbourne, the engineering firm of Geo. W. Kelly & Lewis (KL) was formally established in December 1899 with a £600 capital base. Among a range of products made in the early days, the company was, at one stage or another, involved with producing engines or parts for many famous names in the (vintage) engine business, including Hornsby, Horwood Bagshaw, Bartram, Dalgety, Gippsland & Northern, Dangar Gedye ... and they also claimed to have built the first aeroplane engine in Australia. In the 1930s KL became the Australasian agents for Lanz Bulldog tractors. But German tractor imports ceased abruptly with the outbreak of hostilities. During World War II the Lanz Mannheim Works in Germany was totally devastated.

KL Bulldog

Following the war, Kelly & Lewis established a separate division, KL Tractors Ltd, at Springvale, Victoria, to produce the KL Bulldog tractor although they had no agreement with Lanz to manufacture tractors. Kelly & Lewis had the assistance of the design engineer A. Lou Murr, who joined Lanz in Germany in 1929 and came to Australia in 1937 as manager for Lanz Australia. He was interned during the war, but afterwards decided to settle here and joined KL Tractors Ltd as engineer responsible for the KL tractor project.

The first Lanz Bulldog, Germany 1923.

What was left of the Lanz works in Mannheim after 15 Allied bombing raids in 1945. Note the producer-gas generators on the damaged tractors. (Herrmann, 1987)

The KL Bulldog, announced first in August 1947, was to be identical to the 40 HP Lanz, powered by a two-stroke single-cylinder engine with hot-bulb ignition.

The Springvale works was to start out as an assembly plant, with machined components being built under contract by the government ordnance factory in Bendigo, as part of the Federal government's assistance package. The government was actively encouraging several companies to build tractors locally.

The KL tractor plant was officially opened in March 1949 when, it was stated, 20 tractors a week would be produced 'immediately'. The demand would be '10,000 a year under existing trade conditions' said the optimistic KL chairman, A.L. Bruns. 'Only the ball bearings would be imported, the rest would be Australian-made' (*Power Farming*, April 1949). Machined components would be built under contract by the government ordnance factory in Bendigo and production was anticipated to eventually reach 17,000 tractors a year. Despite these ambitious plans, there were protracted delays in assembling KL Bulldogs at the Springvale plant. Murr's stay with KL was brief and terminated after a disagreement. Serial production only commenced in late 1949. The first KL Bulldog to come off the line was given the serial number

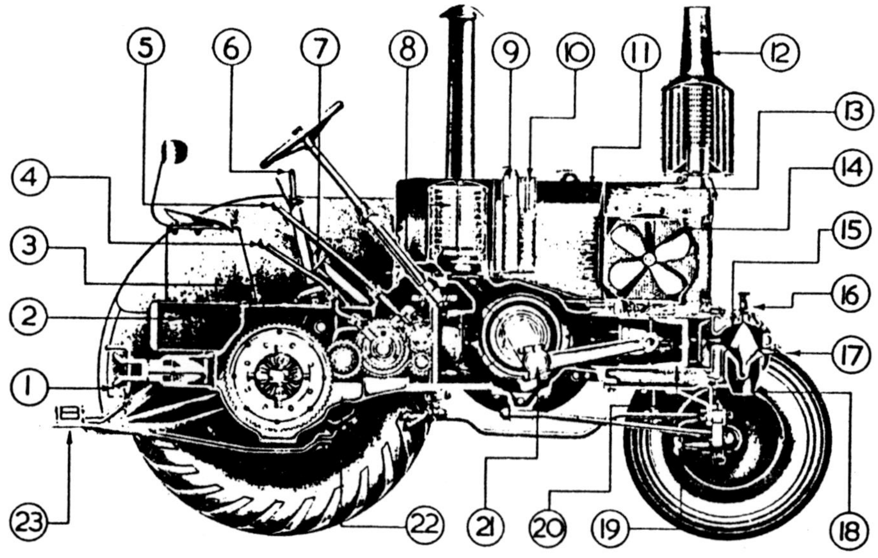

Cross-sectional views of the 1929 Lanz Bulldog and the 1951 KL Bulldog. The Bulldog tractor design was copied in many countries.

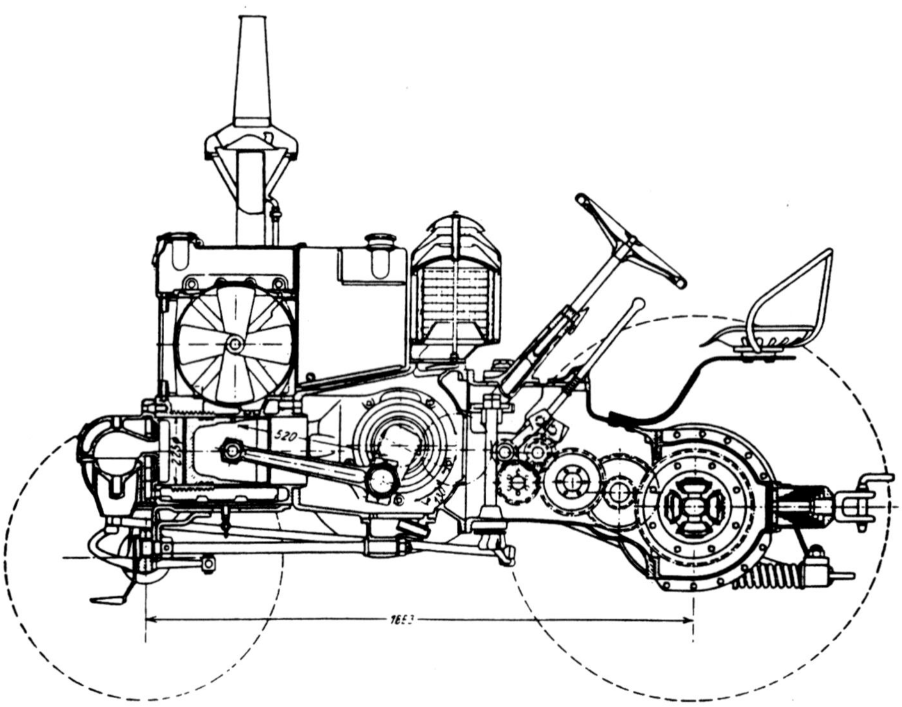

1001, based on the view that farmers would not like to own the first tractors out of the factory. The KL Bulldog was formally tested at Werribee as Test No. 12, March 1950. Maximum belt power was 44.2 HP, and maximum drawbar power 41.0 HP on test. In the event, however, the government ordnance factory was unable to keep up the supply of parts for KL Tractors to meet orders. This, along with other manufacturing difficulties, plagued the company.

Meanwhile in Germany, the Lanz works were rebuilt immediately following the cessation of hostilities. They were back into production by 1948. In 1950, Kelly & Lewis resumed as their Australian distributors. In June 1950 the Commonwealth government requested KL tractors to cease importing the Lanz Model L Bulldog in favour of importing larger Lanz tractors.

The KL Bulldog appears to have been locally produced up to 1954, with the last new one being sold out of inventory in 1960. *Power Farming* advertising shows the company offering KL Bulldogs at cut-price rates in 1951 — perhaps indicative of the fact that competition was displacing old technology. The newer Lanz tractors were by then being offered with electric starters and petrol/ spark ignition to aid startup.

After making around 900 KL Bulldogs, the KL Tractor Company wound down production. Meanwhile, Lanz set up their own network in Australia as Lanz-Australia Pty Ltd in 1953. Then Deere & Company acquired a controlling interest in the German Lanz works in October 1956.

By 1958 Kelly & Lewis faded from the KL and Lanz tractor scene. In 1959 the John Deere-Lanz distribution network was formed in Australia and the tractors were renamed accordingly. But the Lanz tractor and name were phased out entirely by John Deere around 1964.

[An important source of information in this chapter came originally from *Power Farming*, through Peter Willcock of Merbein who compiled a history of the Bulldog in Australia. His findings were published in *The Old Machinery Magazine* through 1986. Gary Vines, Museum of Victoria, compiled information on the KL Bulldog based on interviews with individuals who worked with the Kelly & Lewis company. The assistance of the former Rural Science Section of the Museum of Victoria under Martin Hallett is gratefully acknowledged. Ray Gillett of Ringwood has maintained a KL register.]

*KL Bulldog 40 HP crude oil tractor. It could 'run on sump oil' claimed one advertisement (*Power Farming, *August 1951). The two-stroke engine of the Bulldog Model L had a single piston of nine inches diameter and 10.4 inches stroke. Tractor weight was 7200 pounds.*

This restored Bulldog belongs to Gary Skinner of Guyra, NSW. (Courtesy The Old Machinery Magazine)

18 THE CHAMBERLAIN TRACTOR SAGA

Born in 1908, I have been lucky enough to live through the greatest period of technical achievement in the entire history of the world.
— *A.H. Chamberlain in 1986*

A.H. (Bob) Chamberlain, born 1908 (right), with brother H.F. (Bill) Chamberlain, in 1978.

Bob Chamberlain with the author in Toorak, Victoria in 1988.

In 1986 Mr Alan Hawker (Bob) Chamberlain was contacted by the author for his story on Chamberlain tractors. The request yielded some unique material that had never been published. His story was then published in *Agricultural Engineering Australia* in 1988. Extracts follow, in a slightly abridged form, along with other information on his tractors and on Chamberlain John Deere.

The story begins with Bob Chamberlain's father, Bert Chamberlain (brother-in-law of Harry Hawker the aviator), who had established the Australian Ball Bearing Co. Pty Ltd in Melbourne. In 1927, following technical school and at the age of 19, Bob joined his father's company as an engineering trainee. The three Chamberlains: sons A.H. (Bob), H.F. (Bill) and father A.W. (Bert) actually laid out the design of an Australian farm tractor in the late 1930s. Bob Chamberlain was the prime motivator behind the tractor project, but then the war intervened.

Birth of a Tractor

'I had the idea of building a heavy duty wheel-type tractor before World War II and set out the basic design of a machine to suit the requirements of the wheat farmer.' So said Bob Chamberlain, who spent time working as a tractor mechanic servicing tractors in the field in order to study farmer's requirements at first hand. He concluded that tractors of the period fell short in two areas: their engines required costly servicing at all-too-frequent intervals, and the transmissions did not provide enough ratios to handle all conditions efficiently.

A deliberate effort was made throughout the design period to give these two conditions top priority. The result, after the war, was to be a tractor powered by the most rugged engine in the industry, coupled to a nine-speed transmission, with plenty of gear ratios packed close together in the range where the farmer did most of his work. Half a mile per hour extra around the 4 mph range could equal an extra hour's work per day — if the power were available and the implement could handle the extra speed without excessive soil displacement.

But to return to the 1930s. The automotive parts business grew rapidly during the decade and the Port Melbourne company had more work than it could handle by the outbreak of war. The basic components of the first pilot model tractor were ready before the war, but the tractor was not assembled. Then Bob Chamberlain was posted to the US as a member of small Allied design team to develop a tank to match superior German tank technology. This was an ideal situation for Chamberlain, as it exposed him to the top designers in the US automotive, tractor and aircraft business. Such were the resources at the disposal of the design team that, starting from the first line on the drawing board to volume production of the new M3 tank (built by Chrysler of Detroit), took just nine months.

Bob Chamberlain completed his work in North America and returned to Australia in 1943. At this stage of the war the government was becoming increasingly concerned with the nation's ability to produce enough food. Mechanisation was imperative, but there was a shortage of tractors and parts. Federal economists

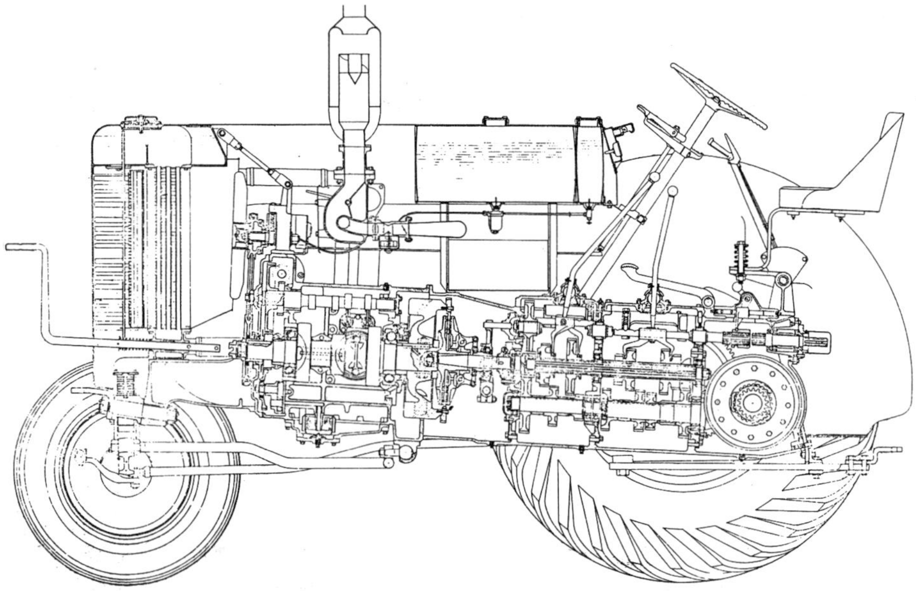

Copy of an original line layout of the 40K, as drawn up by Bob Chamberlain.

By November 1945 the tractors, designated Model 40K, were ready for field testing. An official test report was considered necessary to provide an independent assessment in a similar way to the Nebraska tests on North American tractors.

An approach was made to Melbourne University. There Gilbert Vasey, who later established the country's first Agricultural Engineering Department, set about locating suitable test equipment. The CSIR's Division of Aeronautics (under Ron Wright) had all the equipment needed for engine testing. This included suitable engine dynamometers and apparatus for accurately measuring fuel consumption, so this part of the program was easy. Gilbert Vasey found that the university had an unused recording drawbar dynamometer. All that was needed was to find the means for applying suitable drawbar loads for the range of outdoor tests, which were to include a continuous 10-hour run at 85% full drawbar power. One answer was to haul a group of big trucks linked together and towed in various gears with the fuel supply turned off.

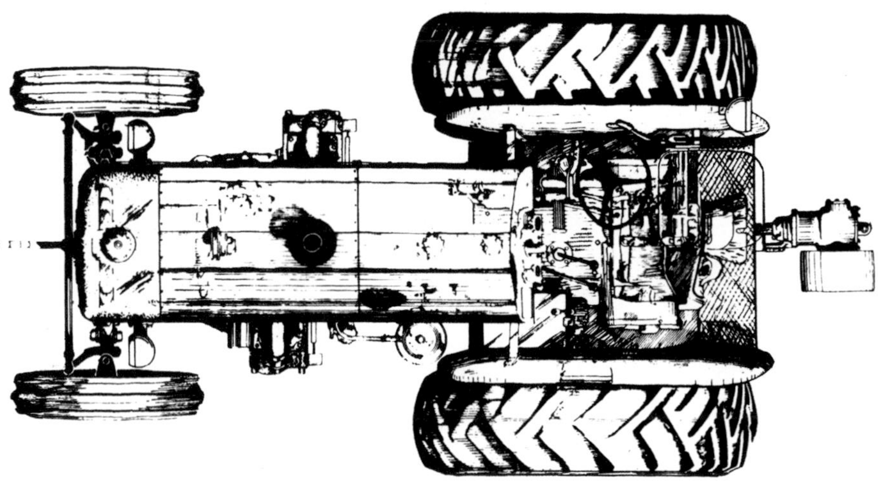

Chamberlain 40K/KA illustration, amended from the tractor manual; some parts ghosted. Belt pulley attachment fitted.

estimated the shortage could last for a decade.

This prediction was a bit off and the tractor shortage was actually taken up by 1952. Nevertheless, Norman Makin, Minister for Industry in the Chifley government, knowing that the Chamberlains had a tractor on the drawing boards, approached the Rolloy Piston Company (owned by the Chamberlain family) to see what was needed to get it into production and how the government could help to get the project under way.

Prototypes Assembled at Last

Two tractors were assembled at the Rolloy plant from the parts that had been built pre-war. The only imported parts were bearings, carburettors and electrical equipment. A 'slave' or test engine was run on town gas on a dynamometer test bed at full power for 24 hours a day for months on end.

Norman Makin, in Canberra, was asked for help in getting the use of some trucks from the government. 'Yes,' he replied, 'I think it can be arranged, but it would take at least several weeks to process the application through several departments.'

Gordon Brookman, a staff engineer, decided to take matters into his own hands. Within an hour Gordon had organised with the Transport Department manager for five Mack trucks. By the following day, the prototype tractor was out on the Albert Park Lake ring road — little used in those days — hauling a row of the big Mack trucks for the drawbar dynamometer tests. The outcome was a set of results which indicated that the 40K was performing well up to the best US tractors of the time.

This was the official Australian Tractor Test Number 1.

Production was the next target, but family finances could not stretch far enough to insure control for the launch of a tractor production venture. Then, seemingly out of the blue, in walked Norman Fernie, director of Industrial Development for the Western Australian government. He said that Western Australia had plenty of advantages if the Chamberlains would set up their new industry over there. The Western Australian government had been promised help by Prime Minister Chifley to set up heavy engineering industry near Perth. This would provide employment for many Western Australians who had returned from the war and didn't want to go back to the land where they would be underemployed.

Western Australia Beckons

Near Perth there was a large new vacant factory that had been constructed for ammunition production. There was a guarantee for the family's overdraft, banking support, and there were some new machine tools at hand that had been sent over for an abortive Australian military tank project.

Weighed against any advantages of setting up production in the west were some foreseeable problems:

- The extra costs compared with the eastern states. A lot of electric power would be required for the foundry and Victoria's Latrobe Valley, with its coal deposits, was producing the cheapest power in Australia. The WA government countered this by giving an initial undertaking to provide power at Melbourne prices.
- The shortage of suitably trained labour.
- Difficulties with the availability of components.
- The costs of distribution to eastern markets. It could cost as much to ship a tractor there as to ship it to England!

The persuasive Premier of Western Australia, Bert Hawke (uncle of Bob Hawke, who later become an

The original team which went to Perth in 1949 to start the tractor production program. Back row: (left to right) Rex May, A.W. (Bert) Chamberlain and Bill Balgarnie. Front row: H.F. (Bill) Chamberlain, Len Elvage, R. Lawrie. Bert Chamberlain was the father of A.H. and H.F Chamberlain.

Australian Prime Minister) offered further promises of support. A massive engineering strike in Melbourne, which lasted six months, helped galvanise the decision. In 1949 a tractor factory was established at Welshpool, WA, as Chamberlain Industries Pty Ltd. That year, with a staff in place of 265, the pilot production run of 10 tractors was completed.

The initial production target was 1000 tractors per annum. By 1951 the 40Ks were being turned out in

Pilot Model 40KA tractor on belt test at ARL in 1946.

The first pilot model Chamberlain 40K tractor was assembled and ready for field testing late in 1945. Beside it is the 1100 cc Chamberlain Special, a supercharged two-stroke front-wheel-drive racing car built by Bob Chamberlain. This car was a successful racer in its day, and still runs at vintage rallies.

quantity and they sold well. Most were purchased by Western Australian farmers who paid cash direct at the factory. Many local purchasers drove their new machines to their farms, as the company had yet to establish a sales organisation and dealer network.

The experience of setting up the manufacturing facility was a salutary one. In the first place, Perth's power supply was initially not up to the demands from the Chamberlain foundry, so the company had to install their own 1400 HP diesel genset for supplementary power. On the commercial side, they faced some disappointments. Wundowie iron was more expensive than BHP's, but the quality was good. Finally, the coastal shipping routes were inadequate and some tractors had to be sent east by road and rail — at much higher rates still.

Happy in the thought that the tractor shortage would continue for several years, the fledgling company went into 1952 hoping to sell 2000 tractors for cash in all Australian states. But 1952 was a disastrous year for machinery sales. That year the government reduced their 40% investment allowance to 25%; Ford in England had fitted the 45 HP Perkins P6 diesel to their tractors and had shipped a large number to Australia. This model sold cheaply and, with Ford Australia's strong and well-established sales organisation, posed an enormous threat to the Chamberlain product.

Company in Jeopardy

The company was committed. They had bought in parts, bearings, forgings, radiators, electricals and steel, and had assured the Western Australian government that they would employ 2000 men. These factors, along with previous financial obligations, practically prevented the impending decision to layoff 1500 people and to cancel contracts. Production went on and unsold tractors began piling up all around the factory, which could produce 60 tractors a week at that time.

The company was in serious jeopardy. None of the Chamberlain family were drawing salaries from the tractor company. Those came from the Rolloy Piston Company. At that time, the Holden motor car was just getting into volume production and the successful automotive component company was in

Side view of the Chamberlain 40KA pilot model outside the factory. The tractor had a four-stroke, two-cylinder engine of 42.7 maximum belt HP. The two models, the 40K and 40KA, differed only in their gear ratios. Just under 2000 of this tractor series were produced up to January 1955, when the last of the series was made.

Some of the 1000 or so tractors which accumulated around the factory with the 1952 market downturn.

Chamberlain Model 60DA tractor, as advertised in Power Farming Technical Annual *for 1953. This tractor was powered by the GM Series 71 diesel engine of 60 HP rating. It was shortly after upgraded to 70 HP and the tractor called the 70DA. The last 70DA was built in 1967, after almost 2000 of this series had been produced.*

high gear supplying components to Holden.

Bill Chamberlain made two critical decisions in order to save the tractor company. First, it had to diversify. They would manufacture farm implements, designed to suit Australian conditions. Second, they would build a diesel tractor. At 70 HP, this would be in a power range well above anything else then on offer.

Diesel Tractor

The Chamberlains found that the GM 3.71 diesel engine matched their tractor transmission and running gear. A market survey indicated that 200 units, at first to be called the 60DA (1953), later the 70DA, could be sold annually on the larger farms, and there would be no competition for the design.

Field day demonstrations of the 70DA drew great excitement. The market was achieved with little advertising. The decision to produce Chamberlain stump-jump disc ploughs and heavy scarifiers was also a winner: in the first year of production (1953) more than £1 million worth of implements were sold. Nevertheless, there were a large number of unsold kerosene tractors on hand — and a bank overdraft of £3.5 million. Then with the 1953 Western Australian elections came a new government, when the McLarty Liberal–Country coalition took over after 14 years of Labor leadership.

In opposition, the Coalition had been 'yapping at the heels' of the Labor government about their involvement with Chamberlain Industries. During a particularly bad period of political infighting before the elections, Labor Premier Bert Hawke sought an assessment of the company's situation. He wanted to continue government support for the company. A man of few words, he said, 'Everybody would like to see better roads and schools, it is far more important that the men have jobs and can take their wages home every week.'

Massey Ferguson Makes a Move

When the McLarty government came into power they were faced with some decisions about the future of the company. The Chamberlain family still owned the company but had a big bank overdraft. The new government wanted to get out of the tractor deal. Massey Ferguson — then in a strong position — was interested in buying the company. In 1954 a meeting was arranged for detailed discussions. Present were Tom Ritchie, managing director of Massey Ferguson Australia, a director who had come over from the head office of Massey Ferguson in Canada, Alex (later Sir Alex) Reid who was then under-treasurer to the Western Australian government, and Bob Chamberlain, who was in Perth at the time.

The directors of Massey Ferguson had taken a suite at the Adelphi Hotel. At a relaxed weekend luncheon, talk turned to the business at hand. Almost immediately the Canadian director showed his hand. 'Well, we need to allow for a lot of money to take the plant to its new location and set it up.' This the first mention of such a move. When South Australia was mentioned, Alex Reid exploded: 'In that case we are all wasting our time. It's a nice day — I would sooner be enjoying a game of golf.' With this he just exchanged a few pleasantries and walked out. It was all over in 10 minutes.

But there was another threat. The new government contained some politicians hostile to the tractor project. Indeed, some were on public record as saying they would withdraw government guarantees to the company's bank financiers. Fortunately, this view was not shared by all members of the new government. Charles (later Sir Charles) Court brought several MPs to Welshpool and they were shown over the plant. Court commented to the other members: 'This will work out successfully. It should continue.'

The McLarty government engaged an eminent consultant to investigate the company's position and make recommendations for the future of a company which was paying interest on a £3.5 million overdraft and employing about 2000 men at a time of serious unemployment. The consultant was to stay with the project as long as required. Sir Edwin Nixon of Melbourne, a top-level financier with widespread interests as director of some leading companies, was appointed. He was a director of one company importing tractors and had ready access to field information on Chamberlain products. He set out a list of the figures he would require on all aspects of the company's finances. After checking these over, he didn't mince words: 'I've checked over the quality of the products and they have an excellent reputation. You have made some mistakes, particularly in continuing to build tractors in 1952 when sales were poor. This has tied up a lot of money and is crippling the company with extra bank interest. You will need to improve some details of your bookkeeping. There is one particularly bad item — a lot of money is written down to cleaning and obviously it has been the practice to charge all otherwise unaccountable money to this item — this is not good enough.' He went on to say that there was nothing basically wrong with the project and 'it is now up to us to make a success of it'.

Bert Chamberlain, who was present, picked up that last remark as a great morale-boaster, as by 'us' Nixon meant that he now regarded himself as a member of a team working for success for the Western Australian enterprise. A good commercial manager was appointed, and Nixon sought to defer interest on part of the overdraft although this was not accomplished. The 1952 tractor carryover was eventually sold, though with heavy trade-in allowances. But profits from sales of the 70DA diesel tractors and implements were still not large enough to justify the operation.

A new tractor of advanced design was needed — a versatile machine for both farming and industrial use that would grab most of the business in the biggest sales area. There was probably just one chance at this. If it failed, the company was finished. Design parameters were outlined, based on the results of an extensive field survey. Questionnaires were sent out to thousands of farmers all over Australia and contacts established with the many firms specialising in industrial equipment applications, such as cranes and front-end loaders. A high percentage of farmers answered the questionnaire. The most important requirement indicated was greater comfort. The second most important was greater fuel tank capacity — enough for a whole day's work. High ground clearance, good transport speed, powerful brakes,

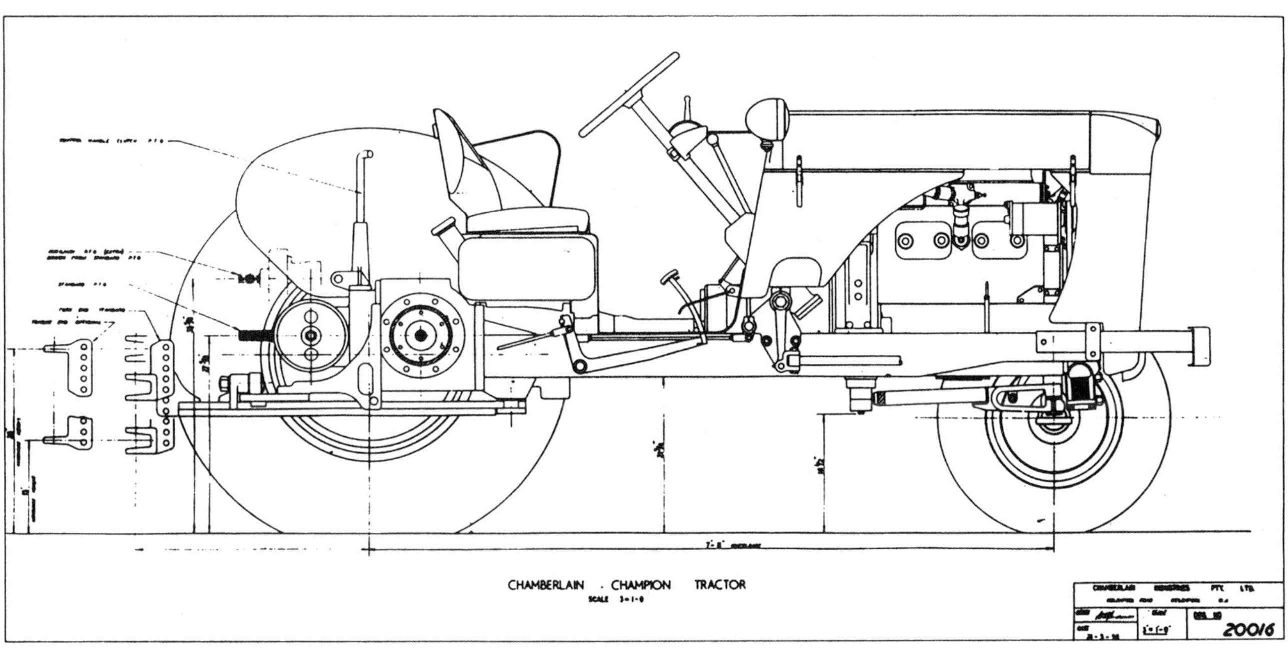

A March 1956 line drawing of the Chamberlain Champion design, which had been conceived in 1954.

and the need for a range of gear ratios followed. The range of Chamberlain tractors already in the field had established such a good reputation that reliability was taken for granted. Bob Chamberlain worked back at the Perth drawing office, mostly in evenings and weekends, to sort out the problems of a new design. By December 1954 he had completed the assembly drawings for the Champion model tractor. These he handed over to Bob Dobson to detail and to design all the tooling necessary for production. The drawing office staff excelled themselves under pressure, as they had earlier in designing the plough. Barry Waghorn designed the transmission and final drives.

Local development of an engine in the size needed was considered unfeasible. Instead, the 45 HP Perkins diesel built in Peterborough was selected. Costing needed to be trimmed to ensure a Federal government bounty and this meant that bought-in parts cost must be kept to a minimum. In the costing of the completed tractor, only £200 could be allowed for the engine and this was much less than the best price Perkins could quote.

Bob Chamberlain made a rush trip to England, where Richard Perkins, the sales director of Perkins Engine Company, was shown costing estimates on the tractor. Said Bob Chamberlain: 'I have authority to order five thousand 45 HP engines but I can pay only £200 each. Get your pencils sharpened and see which parts you will need to leave off to get our cost down to that figure. If it is satisfactory we will make those left-off parts in Australia.' After a few days of discussion relating to the most suitable parts for Australian manufacture, it was agreed to buy the engines less flywheels, oil sumps, generators and exhaust manifolds, for £200 each. The first order for 1000 was placed. Very few engines were delivered to this specification because it was necessary for flywheels and oil sumps to be put on the engines for dynamometer testing at the factory in England. The cost of removing the parts and fitting sheet metal protection in place of the oil sump almost equalled the value of the parts. Final production costs were higher than early estimates. At £220 each, Perkins could supply near-complete engines.

Two pilot model Champion tractors were built. Bob Chamberlain hit on the idea for an unusual advertising stunt to get significant press coverage at low cost.

Tail-end Charlie

The Redex Round-Australia Trial would be run at about the time that two pilot model Champion tractors would be completed, so one of these, with a crew, was offered to the Redex Company to run with the motor trial as a backup and rescue vehicle. Mr Sheppard, the manager of Redex, jumped at the offer. Briefly, the Redex Trial was the brainchild of some members of the New South Wales Car Club. They had wanted somebody to sponsor a trial to run more than 11,000 miles right around Australia, averaging more than 500 miles each day, much of it over rough cattle tracks. Most companies approached had visions of broken cars and dead bodies all over the country and wouldn't have a bar of it. Redex was a small company and agreed to sponsor the trial, little knowing what they were letting themselves in for. Running the 1955 trial strained their resources to the limit, but it was so successful that the publicity gave the impression that Redex was a motor trial. Few realised that Redex was an oil additive.

When the two prototype tractors were being made, one special set of final drive gearing was assembled to provide 65 mph design speed. The special tractor was equipped with a stiff transverse leaf-spring on the front axle, big brakes, and a weight distribution enabling it to handle well at high speed. It was reasoned that with a pair of heavy truck-type telescopic shock absorbers on the front, and rear tyres with truck-type treads, this tractor could handle the toughest conditions. Olympic Tyre Co. agreed to make the purpose-built rear tyres at their own expense, even though this involved making a new mould for a diamond-pattern tread. Perkins Engines agreed to provide one of their field engineers as a crew member to look after the engine.

The Redex Trial was to start in Melbourne, but the tractor was not completed in time. Arrangements were made instead for it to join the trial at Darwin, which meant that it had to run twice over the real horror sections of the trial. The last day was calculated that it could leave Perth to reach Darwin at the same time as the trial cars. The tractor was completed and started up for the first time in the afternoon of the day before departure. Bob Chamberlain, experienced at race driving, continues the story: 'I got on it alone and drove it about 20 miles up the road at around 65 mph. It handled very well and could slide the gravel corners dirt-track style, with no risk of overturning. On returning to the factory I discovered that I had been followed by a police car. The police said that they didn't intend taking any action but couldn't understand what kind of vehicle they were following at 65 mph!'

The tractor left for Darwin next morning with Bill Balgarnie as first driver. As he drove through the factory gateway he said, 'Well, we have to do 550 miles everyday until we get back.'

And that is just what they did. The crew comprised Bill Balgarnie, Bob Patterson from the drawing office, Arthur Osborne from the factory, and Pat Patternall from Perkins Engine Co. A car went with the tractor to provide relief for the drivers, allowing one at a time to sleep as they went along. The car, a Ford, had been specially prepared by fitting an alloy steel plate under the gear linkage and air bags in the coil suspension springs. This same car made all the trial trips accompanying the tractor without trouble, then its pinion bearings collapsed afterwards while being driven at 30 mph on a city street!

The tractor rescued many broken-down cars and sometimes towed two cars at a time for hundreds of miles, at about 50 mph. The following year Redex ran a similar trial and the tractor did the same job, running twice over the worst horror sections. By 1957 the Redex name was associated only with motor trials and all the publicity had overshadowed the company's real reason for being in business. Redex withdrew. Mobilgas next sponsored a similar around-Australia contest with facilities on a much larger scale. That trial was to start and finish in Melbourne. The tractor was dispatched to start with the cars and run the entire distance with them. Before that trial, Tail-end Charlie had already logged more than 24,000 miles and had done 3000 hours of farm work. Running with the 1957 Mobilgas Round-Australia Trial, the tractor covered over 11,000 miles in 19 days.

This tractor is still a runner and is located at the Whiteman's Park Museum in Perth.

Tail-end Charlie, a special-purpose high-speed version of the Chamberlain Champion tractor which completed three major round-Australia motor trials as a backup and rescue vehicle. It was capable of a road speed of 65 mph and could even be pushed to 68 mph (110 km/h).

Dabbling with a Locally Built Diesel

In 1953, it had been judged that there was a demand for a higher powered version of the 40KA, of 55 continuous horsepower, and this was easily achieved by using larger cylinders and related parts on the original 40K engine block. This tractor model was the 55KA, released in 1954. It was sold for five years.

The 70DA tractor had proven an immediate success when it was released in 1955 — but with the GM diesel engine it was a costly machine.

Chamberlains reasoned that they could modify their

Mobilgas display, 1957. One of the first pilot model Chamberlain Champion tractors. It was the new Champion model exhibited at agricultural shows for the first time that had a strong influence on Massey Ferguson's interest in the company. The Champion tractor travelled further than the cars in the rally but finished with the winners.

Right- and left-hand views of the 55DA tractor. The pilot model diesel engine for this tractor was built in Melbourne, production models in Western Australia.

55KA two-cylinder engine, which had a heavy duty crankshaft on large ball races that would be adequate for diesel combustion pressures, and produce a lower cost diesel power unit. But starting such an engine, with its big bore and horizontally opposed piston configuration, would have been a problem. The only practicable way to start this engine with the existing electric starter at the time seemed to be to adopt a system developed by International Harvester Co. for their diesel tractors in the US.

The IH system used an extra valve in the cylinder connecting with an extra displacement combustion chamber in which was installed an intake valve and a spark plug. A lever installed beside the driver's seat sequentially prevented the main intake valve from operating, opened the transfer valve to reduce compression and controlled the fuel pump rack. This Lanova air cell type of combustion chamber was used to reduce peak combustion pressures. Engines with this combustion system were popular in the US, being built then by IH, Continental Motors, Mack Truck, Buda, etc. This design gave smooth running with low peak pressures. Fuel consumption was reasonable, although not as good as with a direct-injection design. So the Australian-built diesel, the 55DA, ran very well when correctly set up and gave pretty good performance. But limited resources put a damper on development. Some troubles plaguing the fuel transfer and injection pump system blunted success. In retrospect, Bob Chamberlain acknowledges that it would have been smarter to have used direct injection in the first place, but the lure of using the cheaper heavy diesel fuels then popular for the older semi-diesels and hot-bulb engines was the diversion.

Since the company already had the Champion model tractor, with its British-built Perkins diesel engine, and the larger GM-powered 70DA, which covered much of the market in 1955, Bob Chamberlain directed that the 55DA be withdrawn from production.

However, things did not go according to plan. Bob Chamberlain was a busy man, juggling business interests overseas as well as an the eastern and western seaboards of Australia. While he was overseas on another trip the company's management committee put the 55DA back into production and ordered in a great inventory of bought-in parts for this tractor model.

Chamberlain Family Out

It did not seem to occur to the government appointees controlling the management committee that the company might have some difficulty selling the 55DA, although the Champion went into full production in 1956 and it became the best seller in its class.

Such were the varying fortunes of a company in which family control was being wrested from their hands by government interference. A fallout was inevitable, even though the company seemed assured of markets with a good product lineup in tractors and tillage equipment. The decision was made by the government in 1956 to break with the Western Australian enterprise.

So the Chamberlain family quit with the west in 1956. They were now free to devote their efforts full time to their Melbourne companies, producing automotive components for a burgeoning market where GM, Ford and Chrysler were taking Australian car production and sales to their best years in the 1960s.

For his outstanding achievements with Australian tractor development, particularly with the Champion, Bob Chamberlain was awarded the 1956 Olympic Medal for Outstanding Design and Manufacture.

The Chamberlain Light Tractor

The tractor market in the smaller size range was dominated in the I950s by Massey Ferguson with their ubiquitous MF35. Bob Chamberlain could not resist the urge to challenge Massey with an Australian-built machine in that tractor size range.

He accordingly set to work to design a light tractor with superior specifications. The engine was to be the 138 cubic inch six-cylinder (3.3 litre) Holden car

Chamberlain Champion Model 9G, 1959–67. Perkins 62 HP at 2000 rpm.

engine, which was already in volume production at low cost — the engine could be purchased complete for just £125 in 1959. Two pilot models of the Chamberlain light tractor were built at the Rolloy Piston Company plant in Port Melbourne — right where the original 40K had been born. One of these, the Holden-powered unit (the other had a M.A.N. diesel engine) was tested at the Australian Tractor Test Station, Werribee. The results were encouraging — it produced 39.6 HP at the belt. But Bob Chamberlain did not carry this tractor development through to production, mainly on the grounds that the government of the time, having rejected the Tariff Board recommendations to support local manufacture, would not now provide the protection by way of import duties which would encourage an Australian-designed tractor of that size.

Back at Welshpool the company was meeting the success which the founders and their adviser, Sir Edwin Nixon, had been seeking. The Champion tractor had become the best seller in its class. It remained in production for 15 years essentially without change. The company went public in 1962 as Chamberlain Holdings Ltd, a company which involved one of Melbourne's leading stockbrokers. Then in 1970 Deere & Co. of Moline bought a 49% interest in the company and it became Chamberlain John Deere (CJD). In the deal, CJD took over the management of John Deere Ltd in Australia.

By 1981 CJD had sold its fifty-thousandth Chamberlain tractor. Tractor derivatives of the Champion, the 80 Sedan Series, were Australia's best selling tractor models for three straight years, 1981 through 1984. The 80B Series tractors were marketed between 1978 and 1984. They were locally designed but fitted with Deere engines, 68 through 119 engine HP.

The 4080 Model (85 HP) was a particularly successful machine, and outsold all others in its power class for five straight years. In 1981 the company had captured 20% of the Australian market. But the next year the farm equipment market collapsed.

With drought and prolonged low commodity prices over the next few years, broadacre farmers almost completely stopped buying machinery in the larger sizes, and many others went out of cropping in favour of livestock production.

Meanwhile the company pushed ahead with the development of the 90 Series tractors — designed in Perth but fitted with US-sourced Deere engines and certain other final drive components. The innovative 700 Series six-rank seed drill, capable of cultivating at one depth and seeding at another, was released in 1985. This followed closely after the launch of the 1051 pull-type harvester, with unique systems of floating feeder house and 'transport cart' which enabled the wide machine to be converted into narrow transport mode in five minutes. The 300 Series scarifier and air-seeder complemented the CJD Australian tillage equipment lineup.

In May 1986, tractor production of the Australian-

The Chamberlain light tractor prototype built in Melbourne in 1956 with the idea of challenging the MF35. This tractor was equipped with a 39 HP Holden engine and was intended for three-point linkage work. This particular tractor prototype belongs to the Museum of Victoria. (Photos provided by Bob Chamberlain)

designed tractor series was terminated in favour of total importation and assembly of tractors from Deere & Co., USA, and of some models from Europe.

[A.H. (Bob) Chamberlain graciously provided much of the early material and many illustrations. Barry Waghorn, long time Chamberlain product engineering manager and Peter Griffiths, managing director, Chamberlain John Deere, also helped in the preparation of this chapter.]

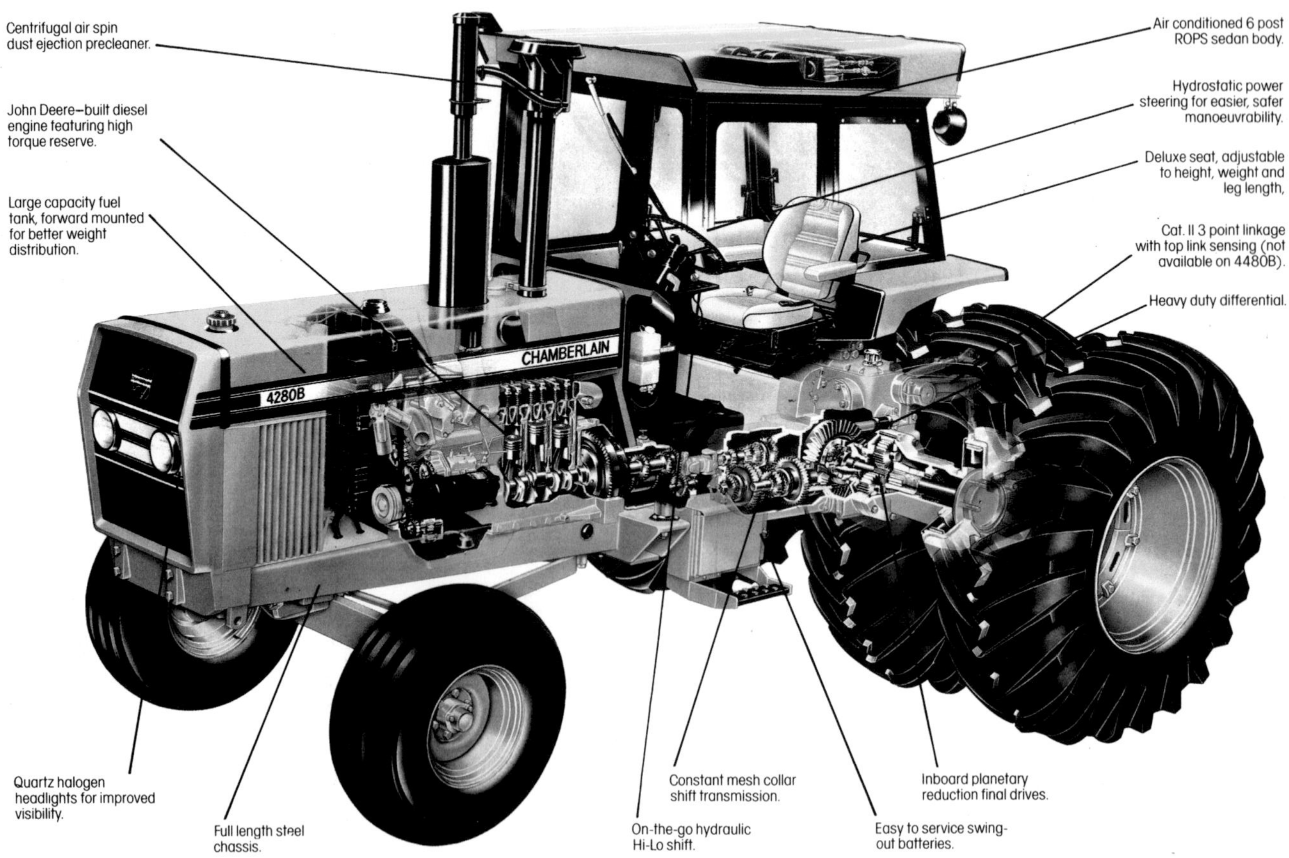

Model 4280B (98 HP) of the Chamberlain Sedan series. This tractor outsold all others in its power class in Australia for five straight years. This model continued the Australian company's two-wheel-drive rugged tractor design style, with 12 forward speeds and four reverse gears; it was fitted with an 85 HP John Deere US-built engine.

ndem Chamberlain rig built by South Australian Mallee farmer, Jeff Morgan. Powered by two six-cylinder GM 671 naturally-aspirated diesels, the machine boasts six · of duals in an articulated chassis with rear axle bogied. This is a working tractor that has been used for crop establishment for 20 years at Karoonda near Lamaroo. Jeff ·rgan states that building this triple-axled tractor from Chamberlain Super 70 basics has cost him less than $12,000 for a one-of-a-kind 450 HP machine. (Courtesy e Old Machinery Magazine, *back cover April/May 2005.) This is the machine on the front cover of the book.*

e Chamberlain light tractor was designed by and two prototypes built under Bob Chamberlain's direction in 1959 with the idea of competing with the Ferguson. e 'Light' was powered by a standard Holden engine. Economic circumstances prevented commercial production. One of these tractors is preserved by the Museum of ctoria. (A.H. Chamberlain provided the photo).

Model 4480 (98 HP) of the Chamberlain Sedan series. This tractor outsold all others in its power class in Australia for five straight years. This model continued the Australian company's two-wheel-drive rugged tractor design style, with 12 forward speeds and four reverse gears; it was fitted with a John Deere US-built engine. It was one of the last to carry the Chamberlain name.

Howard's DH22 tractor was launched around 1930 and was manufactured for 30 years. This DH22 was restored by Ross Sempf, Gatton, Queensland. (Photo courtesy The Old Machinery Magazine, *February 1997)*

19 AUSTRALIAN INTERNATIONAL HARVESTER CO. TRACTORS

The development of this ... Australian-made tractor ... ties together the city and country ... We spent £300 million in the last 10 years or so, importing from overseas tractors and agricultural equipment. We spent this in a way where we did not only lose on exchange, but we did not give the people local employment and so did not help build Australia as we should have. This situation is different when you have a tractor being made in Australia.

— Sir Alan Westerman, CBE, representing the Commonwealth Department of Trade at the official unveiling of the Australian-made A414 diesel at the Twelfth National Field Days, Orange, 1963.

The 1984 takeover of International Harvester Company (IHC) in the US by Case (Tenneco) was a stunning event in the annals of the farm equipment industry. The move takes on some poignancy when it is recalled that early in this century, International Harvester was undisputedly the world's biggest in the farm equipment business. The earlier story of IHC is worth retracing because it had ramifications in Australia.

'The world takes all our output and calls for more', claimed a McCormick Harvesting Machine Company advertisement in 1898. Predecessor to the International Harvester Co., this company formed a massive conglomerate between 1890 and 1907 which at one stage absorbed much of the farm machinery production in North America.

The Harvester Trust

By 1890 the American frontier had about run out. Expanding wheat acreage levelled off, and wheat production was practically static for the first 15 years of the 20th century. IHC farm equipment sales, which had been proliferating, began to decline. The company had became a conglomerate formed by mergers of the major farm equipment companies in the 1890s, which included such previously great names as McCormick, Deering, Champion, Milwaukee, Buckeye and Osborne. This expansion gave IHC control of iron ore leases, steel works, forest reserves, and even a monopoly on twine production for binders.

US field harvesting equipment sales reached peak production in the 1890s at around 152,000 binders and 217,000 mowers annually. The decline which followed the plateauing of wheat production resulted in higher unit overheads in machine production. Furthermore, the elimination of the harvesting bottleneck by these machines highlighted the need for better soil-working tools. Factory space that had been rendered idle by the reduced demand for harvest equipment was converted to the production of other lines. Thus IHC's production and sales activities were spread over a greater part of the year — to include implements, gasoline engines, motor cars — and tractors.

For some years after the mergers, the International Harvester Company had found it expedient to retain subsidiary trade names. McCormick and Deering, for example, had well-established sales agencies in practically every town in the Union during the competitive period of the 1890s. After the mergers, the company could be represented by no less than five dealerships in a given town.

This meant that each dealer had to be provided with complete sets of implements, distinct from each other not only in name but also often by design. The company justified this action in the interests of maintaining the lion's share of the market. But in 1918 this view had to be modified. The US Supreme Court used the Sherman Anti-Trust Act to decree that the company retain no more than one dealer in any US town or city. The company was also ordered to sell off its Osborne, Milwaukee and Champion lines. The number of IHC dealers dropped dramatically as the famous McCormick-Deering marque was adopted. By 1927 IHC's 15-year battle with the courts over anti-trust issues ended. The company saw its market share whittled back from 85% in 1902 to 64% a decade later by such competitors as Deere & Co., Case and Massey-Harris (Quick and Buchele, 1978).

Meanwhile in Australia, in 1906 a savage battle was taking place in the courts and across the pages of the Australian rural press. Chief protagonist was Hugh Victor McKay, pioneer Australian industrialist.

There was some exaggeration in claim and counter-claim but there was also reasonable cause for alarm at the threat posed to Australian interests by IHC. In 1905 the company planned to 'capture the entire trade in agricultural implements' in Australia. McKay was especially incensed by the release of an IHC stripper

During 1905 the international trust was the focus of a pointed and sustained campaign by H.V. McKay in the rural press, especially in the Dairy Farmer and Agricultural News *from Melbourne. Artist Norman Lindsay was commissioned to show the IH octopus in the most unfavourable light in this sampling of cartoons.*

harvester which he claimed was 'a poor imitation' of the Australian Sunshine Harvester, being 'offered by IHC to local farmers at a loss'. It was being dumped on the market, said McKay.

McKay believed free trade to be folly. To him, the only weapon was protectionism. In the same year that IHC launched their first tractor — 1905 — a McKay press statement read: 'A Free Trade policy will increase the population of Chicago and Toronto, but a protectionist policy will increase the population of Australia' (*Dairy Farmer and Agricultural News*, 5 March 1906).

The government finally acted, with tariff protection measures and anti-trust legislation. How the pendulum swings! To get around this new legislation, and assure local product assembly, International Harvester Australia Pty Ltd was established in July 1912. At first, locally built equipment was made under contract. Gaston Bros and Hugh Lennon were two of the companies involved with IHA. In the US, McCormick Deering launched the first commercial power take-off attachment in 1921, and in the following year the famous Farmall Rowcrop tractor series. The Geelong works of the company was set up in 1938. Ambitious plans for local manufacture were interrupted by World War II, but essential war materiel was produced in the new factory in Geelong.

International's first Australian-built tractor rolled out of the Geelong works assembly line in 1949. It was the 34 drawbar HP kerosene-powered McCormick-International AW-6. Just before the war, International's Geelong staff were finding enough things on imported IH tractors that they wanted redesigned or modified to

The first International Harvester tractor was their kerosene-fuelled Mogul, first sold in the US in 1906. The friction-drive single-cylinder engined tractor of 8–16 HP was marketed in Australia from 1909.

*IH's first Australian designed and built tractor was the AW-6, which was released in 1949. (*Power Farming, *October 1952)*

better suit local conditions. Plans were laid to build an Australian model.

After the war, there were enough changes in the design to justify calling the AW-6 an 'Australian-designed' tractor. The whole machine, engine and all, was built in Geelong. The AW-6 was followed by the more powerful Super AW-6, and the Geelong program expanded rapidly to include the local development of headers and broadacre tillage implements. Ten thousand Australian-built tractors were completed by 1953 — in just four years. Eleven tractor models were available in the Australian lineup by 1955. This included their first locally built diesel engine, mounted in the AWD6 which was released in 1954.

Following the AW-7 and the Farmall AM-7 came the launch of the 40 HP A414, which resulted in the doubling of local tractor production in 1963. The A514, A554 and A564 tractor models followed. The last of the Australian-built series were the 756, 766 and 786 tractors, which were sold between 1968 and 1983. Assembly of knockdown tractors shipped from the US (the '66 and '86 series) and Europe followed after 1983, as local design and production was phased out.

IH Australia found itself in dire straits in the early 1980s. They were on the verge of collapse and barely made it through 1982. The company had made a partial recovery by 1985, but at a greatly reduced staffing level. Then in 1986 came the decision to merge the Australian agricultural group with Case, as Case-IH, with headquarters in Sydney. So ended the distinctive IH name as an entity on the Australian farm equipment scene.

*Left: I.H. Australia helps with the war effort. (*Power Farming, *1942)*

The AWD6 used IH Australia's first locally built diesel power unit.

AM-7 — the big row-crop tractor.

The A414 was a 40 HP utility tractor, released in 1963.

The A554, one of the last of the Australian-designed and locally built International Harvester tractors.

20 FORD AUSTRALIA AND THE WALTANNA CONNECTION

Man minus the machine is a slave. Man plus the machine is a free man.
— Henry Ford, 1929

The Remarkable Henry Ford
No discussion about tractors could ever be complete without reference to the remarkable Henry Ford, his great company, and the tractors that bore his name.

Henry Ford, who knew the drudgery of walking many a mile behind a horse-plough and team, started experimenting with tractors in the 1890s, even before the horseless carriage. His farming experiences convinced him of the need for a lightweight power unit that a farmer could afford. He built a few prototype 'dandy' little tractors in the early 1900s, then he sidelined work on tractors until the automobile was in mass production. Once the Model T car was on farms, he reasoned, the tractor would become a necessity. Farmers had been given a taste of power by the automobile. Said Henry: 'The only kind of tractor that I thought worth working on was one that would be light, and so simple that anyone could run it. Also it had to be so cheap that anyone could buy it. With these ends in view we worked for nearly fifteen years on a design and spent some millions of dollars in experiments' (*The Farm Journal*, 1923).

Henry Ford with the Model F which his Australian distributors offered for just £180 in 1924.

In 1915 Ford was upstaged. Word got around that Ford was to release a new tractor, although no formal public announcement had been made. Many farmers were anticipating this tractor, expecting that it would be cheaper than the models and makes then entering the rapidly growing market in the US. They figured that Henry's tractor would fill a place comparable to that occupied by the Model T Ford in the automotive field. 'Taking advantage of this farmer psychology, a group of promoters in Minneapolis, Minnesota, hired a young man by the name of Ford and organised the Ford Tractor Company.' This group was legally entitled to use the Ford name: they had a William Ford as an employee who was not related to Henry Ford (Everett, cited in Gray, 1975). Only a few of these 1915 Ford Model B three-wheeler tractors were made in Minneapolis, but they managed to pre-empt Henry Ford's name on tractors at the time.

The Fordson

Henry Ford's tractor, when in 1917 it was finally released as the Model F, was known as the Fordson. It was among the first mass-produced tractors to use a cast-iron unitary frame construction — a self-supporting framework for the engine, gearbox and rear axle. The unitary frame concept was taken up by most other manufacturers.

Early Fordsons had a worm-gear differential with the worm pinion on the top, where it was inclined to overheat so much that the hot differential housing made the driver's seat unbearable. The worm was placed beneath in the next version.

The demand for US tractors was so great for wartime needs that several companies made adaptations for the T Model Ford and other motorcars so that they could serve as tractors or trucks. A few years after World War I similar variants were offered in Australia.

The first Dearborn-built Fordsons (7000 in all) were rushed into production to meet a British government contract and these were shipped to England to help with wartime food production in the last year of the Great War. In 1918, the 19 HP Model F was offered for sale to US farmers at $750. By 1922, with a new production facility rolling tractors off the line at River Rouge, the price was cut to US$395. Henry Ford was true to his word: 'It is important that it shall be cheap, otherwise the power will not go to all the farms ... Within a few years a farm depending solely on horses and handpower will be as much of a curiosity as a factory run by a treadmill.' He forecast this in 1922. Two years later an advertisement in Australia's *The Farm Journal* shows Henry offering the tractor to farmers for just £180 at a time when comparably sized Australian tractors cost around £500.

This 1915 Minneapolis-built tractor was called the Ford B and a few were made by the unrelated company which set out to pre-empt Henry Ford in the tractor business. He was obliged to name his tractors Fordsons.

One of the local adaptations for the Ford T motor car: the 1922 Tractor Wagon from West's Wheel Works, Sydney.

Production of the Model F peaked in 1923 at 100,000 units, and this total output was reached again in 1925. Almost 750,000 Fordsons were built between 1917 and 1928, the biggest production run ever of any tractor model in the Western world. Fordsons held 75% of all tractors sold in the US, and 50% of the world market. The biggest single customer, strangely enough, was the Russian government, which purchased some 25,000 Fordsons by 1926. In 1927 it was said that 85% of all Russian tractors and trucks were built by Ford (Williams, 1978).

In 1929, Fordson tractor production was closed down in the US, to be transferred to Cork in the Republic of Ireland. That quashed exports to Australia for a time. Henry Ford's choice of Cork for his British base was partly sentimental. His father had sailed from Cork to start a new life in the US after the Irish potato famine of 1846. Henry Ford was probably pleased to be able to use some of his influence to provide employment in that area (Williams, 1978). Many 'Cork' Fordsons — the Model N of 25 HP — found their way to Australia. Fordsons were the prime mover for a number of innovative Australian self-propelled machines and were fitted with all kinds of locally designed after-market attachments, such as improved traction wheel-grips, power take-offs and rotary hoes.

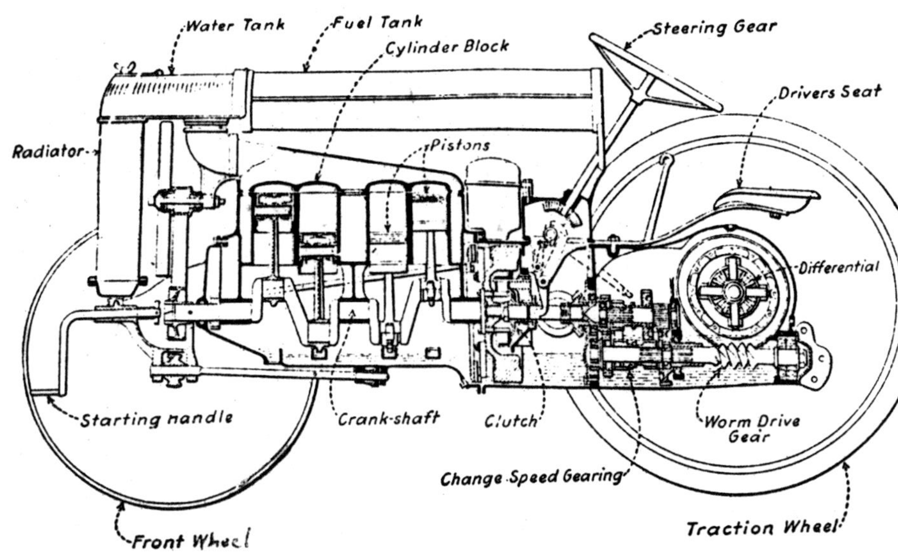

Fordson Model F 19 HP tractor. The Model F was manufactured in Michigan in the greatest numbers of all time for a given model, 100 000 a year.

The Australian-designed Ford 8401, 109.2 engine HP tractor. This represented Ford Australia's first and only entry into the local manufacture of tractors.

Because they could rear up backwards and overturn if misused on towing work, a safety drawbar was designed and made for the Fordson by Norman Wicks of Hahndorf, SA, around 1932. The Wicks family claimed that this patented design was later emulated in the US.

Down through the years, Fordsons and later Ford tractors were widely offered through an extensive distribution network, but were never manufactured in Australia. Until 1980, that is.

The Ford 8401 Tractor

In 1980 Ford Australia launched their Model 8401. This was an Australian-designed and locally built Ford tractor. 'Ford Australia's response to the demands of the nation's broadacre farmers,' said the company brochure. This 109.2 HP two-wheel-drive tractor was designed around the de-rated six-cylinder Ford TW20 diesel engine, with easy servicing and comfort for long hours of operation as design determinants. The Model 8401 was marketed in Australia for five years. A total of 720 units was made.

Ford-Waltanna

In 1986 Ford entered into agreements with James Nagorcka's Waltanna Company in Victoria to complement their imported tractor design lineup with the launch of the FW25 and FW35 tractors. These were four-wheel-drive tractors of articulated frame design, 163 and 195 engine HP respectively. Shortly afterwards, Waltanna's larger models were added — the FW310 (310 engine HP) and FW400 (400 engine HP), then expanded to five models with FW280, FW335 and FW375 models. The Ford-Waltanna arrangement was short-lived, however. In 1988 Versatile was taken over by Ford-New Holland, and the Canadian-built Versatile line started to appear under the Ford blue livery.

James Nagorcka, Founder of Waltanna

James Nagorcka never started out with the intention of developing a tractor business. He worked on his family's mixed grazing-cropping property at Tarrington near Portland, on beautiful rolling country close to the south-west coast of Victoria. But he was a farmer with a very definite bent for machinery. He needed more power to break up land to plant oilseed crops and decided to build his own four-wheel-drive tractor. That was in 1973. The finished product looked good, so he decided to display it at the 1977 Wimmera Machinery Field Days. A Mallee farmer there decided he wanted it and offered to buy it on the spot. A price of around $30,000 was agreed on and Jim Nagorcka was so elated with the deal that he signed a contract with the farmer, but overlooked payment — he had to be reminded by the farmer to collect his cheque! Jim Nagorcka then hired a tradesman and they set about making a few more tractors to order, selling them direct from the works. Production built up steadily until they could produce 20 custom-built tractors a year from the Portland works. Caterpillar engines were used initially. Waltanna tractors were component-built quality machines.

The series was extended to include seven tractors, up to the 55-525 PC, with 525 engine HP Cummins diesel in 1985. Then in February 1986 Waltanna entered a contract with Ford Australia to manufacture Ford's FW25 and FW35 series four-wheel-drive tractors, with Ford 401 engines. On the larger series, the FW310 and the FW400, Cat's 3406B engine was used. The FW Series tractors were designed with 60/40 static weight distribution (front/rear), easy accessibility for service work, ruggedness, and a strong emphasis on backup service to suit Waltanna's discerning buyers. For a time, the line leader of the Waltanna tractor series was the FW525, then a new series with positive-drive rubber-track ground drive was launched. After significant contributions to rubber-track designed tractors with Deere in Waterloo, Iowa, and then a short-lived arrangement with Marshall in England with the rubber-tracked machines, tractor-making at Tarrington ceased in 1991.

Waltanna Model 44-400 made by Victorian tractor builder, James Nagorcka (right). It was fitted with a Gason cab in 1985.

Nagorcka's rubber-tracked tractor had a positive drive system in 1990. Tractor making at Tarrington stopped in 1991 after a short-lived arrangement with Marshall in England. Jim Nagorcka now works on project-specific research activities.

The FW400 (400 engine HP), was manufactured by Waltanna for Ford Australia's product lineup.

21 WILLIAMES HI-TECH TRACTORS AND EQUIPMENT

The Williames Supercat can turn steep, previously inaccessible country into profit-earning land — safely.
— *Williames company literature*

Geoffrey A. Williames, company founder and director, at Warragul, Victoria, with a Supercat in 1988.

In his early farming days Geoff Williames, rural contractor from Warragul, in Gippsland, Victoria, was involved in several near misses while driving conventional tractors on slopes. This served as the original motivation for him to develop his first tractor, the Wildcat. Williames also witnessed a fatal accident when another operator

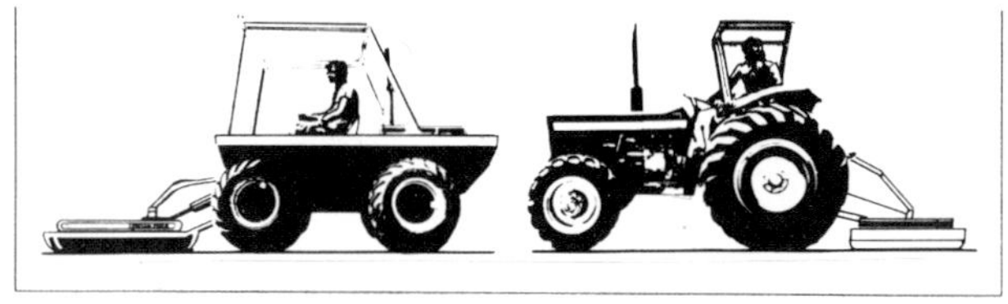

Operator driving positions compared: Wildcat versus conventional tractor.

in the district rolled an old Fordson. He was aware of other such fatalities and, being conscious of the accident potential of wheeled tractors, he set out to build a low profile tractor for the coastal slopes — a design that would overcome the instability of two-wheel-drive conventional tractor designs.

The Wildcat prototype was first tested in 1971 on land close by the operations of the company, G.A. & H.E. Williames, Agricultural Contractors and Engineers, near Warragul.

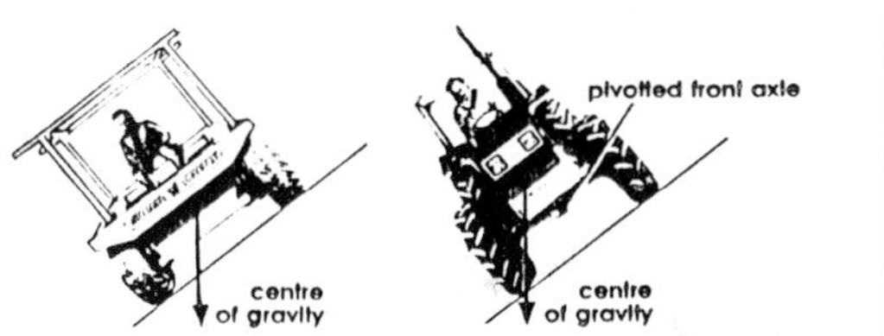

Side-hill stability of the Wildcat was considerably better than for conventional wheeled tractors. The Williames Wildcat four-wheel-drive all-terrain tractor was produced between 1973 and 1986 with a 30 HP Wisconsin petrol engine, although production was interrupted in a lean economic period.

The low-profile tractor design was mounted on equal sized drive wheels and used a four-wheel constant drive, hydrostatic transmission. A new company, G.A.W. Design & Construction Pty Ltd, was set up in 1972 to build tractors. An 18 HP Wankel rotary engine was first used as the power unit. Twenty Wankel-powered Wildcats were built to test the market. Then the 30 HP V4 Wisconsin engine was introduced in July 1973 to give the tractor the extra power that customers were wanting. By the end of two years' production, 180 units had been sold. A range of purpose-built equipment was developed by the company

Williames Hi-Tech International in Warragul, Victoria, made this track-mounted self-propelled tea harvesting machine. Nowadays, the T750 eight-wheel-drive model is offered to order.

for the Wildcat, including front-mounted slashers (hydraulically driven), a back-hoe, and tillage implements. The Wildcat became Australia's first commercial tractor to utilise front-mounted equipment.

Just as the program was starting to gain momentum, along came the severe economic downturn of the Whitlam era. The company was forced to foreclose. However, personnel handling the Wildcat in Queensland perceived a market in that state and began competitive fabrication and sales of the concept in the north. Geoff Williames maintained service for the existing Wildcat tractors, then decided to form a new company. In 1975 he recommenced building specialised farm and horticultural machinery for high-value crops — transplanters, tea harvesters and related tractors.

The company was renamed Williames Hi-Tech International Pty Ltd and set up in a new factory in Warragul as a short-run flexible and innovative business. They offered a tractor — the Supercat — and an associated equipment range built to order. The Supercat was powered by a 43 HP Kubota diesel engine. The all-terrain machine continued the Williames concept of low centre of gravity, manoeuvrability and capacity to work with front and/or rear-mounted equipment. The hydrostatic transmission — no clutch, no gear lever — offered infinitely variable speed.

Williames Self-propelled Tea Harvester and Tracked Prime Mover

The first Williames tea harvesting machine was developed early in 1979. After experimenting with different ground-drives, Geoff Williames opted for a full crawler flexible track design. The basic silent track T5000 tractor is used as a tea harvester or for other operations. The tea harvester module is readily removed from quick-to-locate attachment points. The tracked machine can be converted into a tea plantation track marker, or used as a pruner, or for fertilising and spraying operations.

The T1000 Magic Carpet tea harvester does not run on wheels or tracks but instead conveys itself literally and gently on the tea bush canopy. Manufactured from lightweight materials, it is nonetheless strong enough to be able to perform reliably in arduous conditions on tea estates of all sizes. The idea of floating on the bushes came to Geoff one time while overseas. Taking a break, he lay down on the nearest tea row canopy and had one of those Eureka moments when he saw how easily the canopy could carry his weight. He subsequently patented the idea of a machine that moves itself by conveyor belts, one each side for steering. One version has an articulated frame.

The T1000 can harvest up to one hectare per hour and runs with one operator. It has a floating head for even, accurate cutting, resulting in tea of better quality. Compared to other mechanical harvesting methods, the T1000 requires less labour, cuts with higher accuracy and only needs 18 HP to power via hydraulics. Advantages are:

- Plant density can be increased by 15%, with a continuous tea canopy to be planted without any of the rows and space required for wheels or tracks to run down.
- Two operators, each on six-hour shifts, can harvest in one day as much as 30 operators running 10 hand-held machines.
- The quality of the tea harvested with the T1000 is better due to the high accuracy and uniformity of cut achieved.
- Running and maintenance costs are considerably less than for a fleet of hand-held machines required to do the same job.

Williames has supplied tea harvesting machinery to plantations in South Africa, Uganda, Tanzania, Mozambique, India, Malaysia and Papua New Guinea. This new model was developed based on 15 years' experience and in response to the need for tea estates to increase their profitability.

The silent track on the Williames tractor used a steel wire rope-reinforced flexible urethane belt with bolted-on steel grousers for the ground drive system. Track width was adjustable between 8 to 11 ft. Forward speed capability was 6 mph. The machine has harvested 15.4 tonnes of tea in an 8-hour day.

Williames Transplanting and Horticultural Equipment

The Williames company makes a plant nursery production line for horticultural seedlings, based around the Hexpack growing tray system. The seedlings are transplanted by field transplanters specifically matched to the system. There is a linkage, a drawbar and a self-propelled version. The self-propelled version carries 14 planting heads and is capable of transplanting 55,000 plants an hour with two operators.

Williames Magic Carpet tea harvester weighs just 750 pounds and is powered by an 18 HP engine.

Williames Hi-Tech International at Warragul offers an automated Hexpack transplanting system. The Hexpack tray (left) is fed through planter heads on transplanting machines. The three-point linkage version is illustrated. This unit has three transplanter heads, capable of transplanting 11,000 seedlings an hour.

22 TOFT AND AUSTRAK'S FOUR-WHEEL-DRIVERS

Today there is no room for mistake ...[in our day] we could buy land at a fixed 6% rate of interest and plan ahead.
— *Glen Lamshed, Yorke Peninsula farmer reminiscing on wheat farming in the 1930s.*

Toft R1000

A former member of Geoff Williames' sales team at Warragul saw an opportunity to promote the low-profile tractor concept in Queensland. After some delicate negotiations, Toft Bros at Bundaberg decided to manufacture a higher powered version of the Williames design, which in 1975 they called the R1000. Twenty units of the 53 HP VM diesel air-cooled engined machine were built over a period of several years before Toft decided to drop the R1000 and concentrate their efforts exclusively on the sugarcane machinery business.

The four-wheel-drive Toft R1000, in common with the Williames Wildcat tractor, was hydrostatically driven and capable of delivering traction when only one wheel was on the ground on that side. A roll cage was integral with the main frame, and both seats in the cabin were fitted with seatbelts for rough terrain work. Marketing ceased around 1981.

A 1979 Power Farming *advertisement for Toft's R1000 tractor.*

Austrak's hydrostatic transmission in schematic.

The Austrak

The Austrak tractor also evolved from the Williames Wildcat. The Austrak was first made in Melbourne by Australian Tractor Industries Pty Ltd around 1978. It too was powered by the 53 HP VM air-cooled diesel.

In 1982 the company name became Austrak Industries Pty Ltd, based in Yarra Junction. Then in 1986 a deal was struck between Austrak and A.F. Gason Pty Ltd. The Ararat tractor cabin pioneer and tillage implement maker would produce the tractor for Austrak Corporation Pty Ltd, Ararat, a company in which Gason's would have a 20% equity. This venture was short-lived, however, and was terminated in 1987. Next the Austrak was offered from Austrak International in Knoxfield, a suburb of Melbourne. The basic tractor design was maintained

The Austrak Model RT45 appeared between 1978 and 1986.

The 1986 Gason-built Austrak.

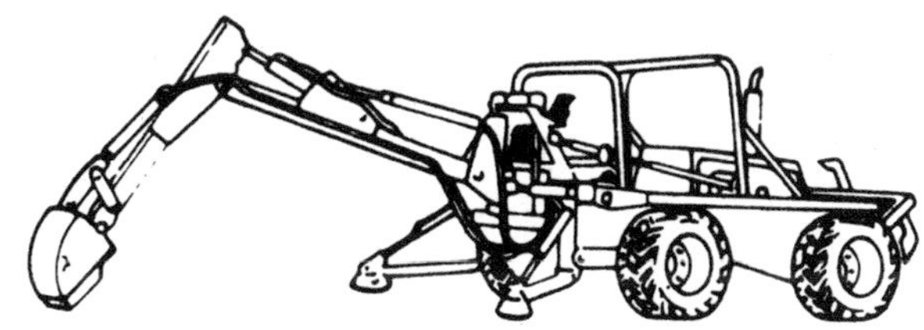

An Austrak equipped with front-end earthmoving equipment was designed to meet a market need between the Bobcat and heavier industrial tractors. The 1988 model was equipped with 56 HP Cummins diesel engine.

throughout the company's peregrinations. The aim had been to slot into a market niche between the Bobcat and the heavier duty industrial tractors. Their latest version was powered with a Cummins 56 HP diesel and fitted with a range of front-mounted earthmoving equipment. Plans were laid to make 100 Austraks a year at the Knoxfield facility, but they never came to fruition.

In all likelihood the emergence of US-built all-terrain vehicles (ATVs) had much to do with the demise of these local ventures.

23 UPTON'S BIG 2WDS — TRACTORS BUILT LIKE TANKS

Having built the world's biggest two-wheel-drive tractors, Carl Upton had proved his point.

The sleepy little town of Corowa on the Murray in southern New South Wales is well off the beaten track, and not a place where you would expect to find a tractor manufacturer. Nonetheless, Corowa is the town where Arthur Upton and his brothers had their general engineering business. The brothers acquired a collection of ex-disposals army vehicles, tanks and spares, during and just after World War II and decided to use these to make some tractor conversions. Perhaps the fact that there was a huge military disposals centre nearby had a bearing on the business.

The first Upton tractors were simply tanks with the turrets and armour removed. These were used — hardly surprisingly — as scrub-clearing machines.

The original Heavy Harry at work. The first public demonstration of Heavy Harry, Upton's original tractor (top), took place at Urana in 1967.

Heavy Harry

Then in the early 1960s they built Heavy Harry — an agricultural-type two-wheel-drive tractor. They took one of the Japanese UD diesel engines that their business was distributing, and teamed it up with a tank transmission and final drive. The 170 HP two-stroke engine (4.9 litre four-cylinder) was Rootes-blower supercharged. Interposed between this engine, with its single-plate foot-operated clutch, and the tank drive shaft, was a four-speed primary gearbox. Then came a five-speed Mack-built tank gearbox, differential and axles — a transmission designed for tanks to withstand 450 HP. Thus the UD4 tractor inherited 20 forward gears, with a speed range from 0.5 to 55 mph.

While Heavy Harry was being developed, Carl Upton started to work for the firm and the team decided to get serious about selling tractors. Their first tractor sold for $4000. It went to Springshaw, and was driven the 1200 miles from Corowa to its Queensland home by its new owner. That took three days at an average speed of 44 mph. The driver of the Holden car who brought the purchaser down to Corowa could barely keep up with the tractor. But at least the Holden driver didn't freeze out in the open, as did the tractor driver at night! Cabins were to come later on Upton tractors.

Later variants of the first Upton tractor series were fitted with an extra gear or overdrive in the primary gearbox, giving 25 forward speeds with a top speed of 70 mph — 'enough to burn off the Chamberlain Countryman any day' gently chided the designer, who may have wanted to match his machine with Chamberlain's Tail-end Charlie.

The third version of the 180 HP Upton tractor.

In 1969 the Model 225 Upton tractor was completed. It was equipped with the Nissan UD225 HP five-cylinder engine, of 6.2 litres displacement. The same 25-speed gearbox and final-drive configuration was exploited. But there were limitations with those tank final drives, one being their limited availability and another their vulnerability to axle overloading by dual wheels: a situation that was never anticipated when merely 'torqueing' the drive sprockets on tank tracks. The tank axles were not designed for the additional heavy fluctuating bending loads that the wide-axle dual wheels imposed.

In 1976 the first of the MT-Series Upton tractors was built. It was initially fitted with a Hercules six-cylinder 200 HP engine, but a few months later it was repowered with a Cummins NTC-290 (290 HP). Still a two-wheel-drive tractor, the MT sold for $48,000 in 1976. This tractor series used commercial planetary final drives and axles, and it was fitted with a well-designed and stylish cabin.

With the MT-Series, Carl Upton's vision of building the world's biggest two-wheel-drive tractors was beginning to unfold. He left off using army tank final drives in favour of truck-type transmissions, for which spares and service were more readily obtainable. The Spicer SST-14 14-forward speed, 4-reverse transmission was adapted to ZF planetary hub final drives with ZF outboard brakes. The MT tractor series were custom-built machines, designed to take any 200–300 HP engine — Scania, Cummins, GM, Mercedes or Caterpillar. All-up weight was around 24,300 pounds, or 31,450 ballasted. With a 200-gallon tank, the tractor could work day and night between fills. The fuel tank illustrates the deliberate way that Carl Upton went about building weight into the tractor: the tank on the MT855 weighed 1½ tons — empty! It was built of $^{5}/_{8}$ inch thick steel plate. The mudguards were fabricated out of one-inch steel plate. Carl Upton was always quick to point out that his combination of Scania V8 diesel and ZF final drive was selected in advance of J I Case, who in later four-wheel-drive tractors used a similar combination. A 1977 comparison of the Upton MT855 with the Case 2870 follows:

The Model 225 Upton tractor was offered in 1969 with UD 225 HP supercharged engine, and rubber torsion block front axle suspension.

Left: Carl Upton with his Model MT855 two-wheel-drive, 200 HP tractor in 1976. Carl was involved in designing tractors just after he left technical school nine years earlier. Right: At the 1977 Henty Field Days, Upton's MT-855 sported a V8 Scania 300 HP engine. Standing with the tractor are Wayne Smith (Alfarm representative), Carl Upton, and Peter Kerr, owner of this tractor. The two-wheel-drive tractor was said to match the performance of the 4WD Steiger Panther of 310 HP at Henty.

	Upton MT855 two-wheel-drive	Case 2870 four-wheel-drive
All-up weight (ballasted)	31,450 lb (14,295 kg)	28,500 lb (12,927 kg)
Engine	Scania V8 300 HP	Scania 6-cylinder 300 HP
Cost	$53,000 with rear duals	$64,690 on singles all round
Origin	Corowa, NSW	Wisconsin, USA

This was almost half of the cost per HP of the US-built Big Bud, which had a weight-to-power ratio of 133 lb/HP at that time.

The downturn of the rural economy in the 1980s, coupled with a need to concentrate on the mainstream product lines of the company — irrigation equipment — caused Carl Upton to delay proceeding to any larger sized tractors or embark on four-wheel-drive versions. Having built the world's biggest two-wheel-drive tractors, Carl Upton had proved his point.

[After going through several hands and a sheriff's sale, the original HT-14/350 is on the Bowden farm in South Australia, where it is being restored. There were also several heavy-duty two-wheel-drive tractors made by Yewers in Bellevue, WA, in the late 1980s.]

The Upton HT-14/350 tractor made its debut at Agquip Field Days in 1978. This tractor, with 350 HP engine and 14-speed transmission, was and still may be the world's largest two-wheel-drive tractor. Carl Upton, designer, and George Tobias, fabricator, stand beside the tractor from Down Under that challenged the world.

World's Biggest Two-wheel-drive

The last round in the Upton tractor saga was the Upton HT series, launched at Agquip Field Days in August 1978. The HT-14/350 — the biggest yet designed by Carl Upton — was supplied with a 350 HP engine (engine make optional). The transmission was a 14-speed Spicer gearbox and Eaton drivehead into an Australian-built final drive. The tyres were 20 ply 33.5 x 33 earthmoving singles, which produced a ground pressure of around 20 psi. Again, the Upton philosophy of building weight into the components to get the weight onto that drive axle was followed.

On the HT-14/350, the base of the 350-gallon fuel tank was $2^3/_8$-inch thick plate and the rest of one-inch steel plate. It weighed three tons, empty. The drawbar was constructed of 6 x 4 bar springsteel and was built within a 4-inch plate frame. The pull was taken by a hardened 3-inch diameter hitch pin! The tractor was designed to provide 147 pounds of tractor weight for each engine horsepower, which at the $70,000 price tag in 1978 was exceptionally good value ($200/HP) compared with any of the four-wheel-drive tractors available at the time.

Upton concept 4WD tractor. Carl Upton was prepared to consider building 4WD tractors, even up to 600 HP, had the market been more attractive.

At Agquip 79 Upton's HT-14/350 hauled a 60-foot blade plough (11 modules) at 5 mph. A special feature on Upton's two-wheel-drive Model HT-14/350 tractor was the swivelling seat, capable of turning 180 degrees. The seat module included the throttle control and brake valve, which moved around with the operator. This tractor's drawbar alone weighed 630 pounds and was constructed from 6 x 4 inch spring steel (right).

Waltanna Model 44-400 made by Victorian tractor builder, James Nagorcka. It was fitted with a Gason cab in 1985.

The Australian-built Ford FW25 four-wheel-drive tractor which first appeared in 1986 (163 engine HP). It was built by Waltanna at Tarrington, Victoria, for Ford Australia. (Author's collection)

Geoffrey A. Williames at Warragul, Victoria, launched a series of low-CG 4WD machines initially as the Supercat from 1973 on, and was followed by others who came out with similar concept designs. The Austrak appeared between 1978 and 1986.

Williames Hi-Tech International in Warragul, Victoria, made this track-mounted self-propelled tea harvesting machine with locally-built silent track — a wire rope-reinforced urethane belt with bolted-on grousers for traction. It could harvest 15 tons of tea leaf from 30 acres in a day. Nowadays, the T750, an eight-wheel-drive model, is offered to order.

Baldwin Model DM600 broadacre rig. This four-wheel-drive 600 HP machine was one of the Baldwin tractor series produced in Sydney between 1979 and 1988.

Horwood Bagshaw's Acremaster 4WD between 1986/87.

Richter Self-propelling Palm-Fruit Tree Harvester.

Richter Hydrostatic Tool Carrier Model TC22 at work in vegetable transplanting mode. (Source: Richter photos)

24 SEMPF, FREEMAN AND AION — TRACTORS FROM THE DOWNS

It needed a lot of labour on the farm in the horse days. Nowadays one man with modern machinery can work the equivalent of 10 farms in those days.

— *Wilbur Howcroft, in* Boots and All, *1984.*

Ces Sempf

Ces Sempf with spud skeleton wheel of the type which he manufactured for certain makes of row-crop tractors.

It was once said of Ces Sempf that he was 'born ten years too late'. At least that's what was said about him when he had an accident on the farm — 'if he had been born ten years earlier, he would have had more sense'. He didn't mind telling this against himself because the accident led him into building farm equipment, turning a hobby into a manufacturing business.

C.H. Sempf was born and reared in the Lockyer Valley of Queensland. He farmed there and in the South Burnett district, and in his spare time built or modified farm machines. While tree-clearing he suffered the severe back injury that forced him to change course.

In 1957, Sempf started making market-garden implements and modifying existing equipment for special applications in the Brisbane suburb where he had a backyard workshop. From this he started making walk-behind garden tractors, and built his prototype Rowvision tractors. In 1962 he moved the business to Gatton, the main town in the prosperous onion, potato and vegetable-growing area of the Lockyer Valley.

Semp Rowvision 80 row-crop tractors, variants and attachments were manufactured by C.H. Sempf & Co. between 1959 and 1983. They used 7 or 8 HP Briggs & Stratton engines, 5.50 x 16 rear tyres, 4.00 x 8 front tyres, and could be equipped with mid- or rear-mounted gear including harrow (above), four-row planter (above right) or boom spray (lower right).

The first Semp Rowvision tractor was a tricycle row-crop machine. The earliest of the Rowvision series was powered by a 7 HP Briggs & Stratton engine. As time went on the units were fitted with bigger engines and twin gearboxes to provide 12 forward speeds and four reverse gears. Price for the four-row 8 HP machine for onion cultivation was £595 in 1963. Tractors were manufactured at Gatton for the next 20 years. In their heyday, the Sempf works could deliver a custom-built tractor six weeks after order.

While tractors became the line leader, Sempf's engineering enterprise included such products as: a range of implements for the row-crop tractors; spud skeleton wheels made for attachment to popular makes of farm tractors; boom sprayer attachments; a high-clearance self-propelled boom sprayer; and special-purpose self-propelled transplanters with tandem drive wheels, in four- and seven-row versions.

After making about 100 tractors, Ces Sempf found the business a good deal more than a retirement job, and he bowed out gracefully around 1983.

[Cecil Sempf of Gatton provided the information at first hand for this segment. All the photos here are from his personal collection.]

Sempf high-clearance row-crop sprayer, with single front-wheel-drive from Wisconsin 21 HP engine; 6 foot clearance; wheel track adjustable 6–9 feet. Driver and engine swivelled with front axle when turning.

Freeman's Engineering

Also in the vicinity of the Darling Downs there was a Freeman tractor made around the mid-1970s by Freeman's Engineering of Toowoomba. Few details have been found, apart from those in the caption below. It is not known how many were built.

What makes this tractor interesting is that the Freeman articulated 4WD tractor was fitted with two GM diesel power plants providing 213 HP to a master gearbox. All-up weight was 29,120 pounds, ballasted on 23.1 x 34 tyres all round.

[Information provided by Peter Bullock, Lismore]

Freeman articulated 4WD tractor that was fitted with two GM diesel power plants.

Darcy Bischoff's Aion Tractors

Aion Trading Company was the name of Darcy Bischoff's heavy engineering works at Dalby. The Bischoff business was geared to produce heavy equipment, machine tools, implements and — for a time — tractors.

The eight two-wheel-drive tractors that Bischoff built were made between 1967 and 1981, the first of them for use on the family property. The rest were custom built to clients' needs, or as prime movers for grid rollers or cranes.

Darcy Bischoff used General Grant or Lee tank transmissions, consisting of two Mack gearboxes to give 30 forward speeds, including the 60 mph road gear. The first units were powered by low-revving Gardner 6 LKB diesel engines of 180 HP. Fully ballasted tractor weight was around 15 tons and cost was $16 000.

One application of the Aion tractor was for use as a prime mover for grid roller work. Power was 180 HP Gardner diesel. All-up weight 26½ tons, unballasted. Shown with grid roller raised and lowered at the Dalby works.

Aion Trading Co. Pty Ltd, Dalby, were makers of this Aion two-wheel-drive tractor in March 1971. The 14-ton tractor was 20 feet long and cost $16,000.

The most powerful model was powered by a GM 225 HP two-stroke diesel. The Aion tractor had a sprung front axle to improve ride. They were built to last and some tractors logged 20 000 hours of satisfactory use.

Walker's Staghound

Peter Walker of Bowenville, Queensland, originally built this Staghound, said to be first 4WD tractor used on the Downs, out of war surplus equipment c. 1953. The original Staghounds were military armoured cars. Repowered by two Leyland diesels at about 300 combined HP, this tractor could pull 30 feet of chisel ploughs or three HB combines. The unit can now be seen displayed at Highfields Pioneer Village. [Information provided by Ray Ashford at Highfields.]

25 THE BIG FOUR-WHEEL-DRIVE TRACTOR BUILDERS

Agricultural tractors should be designed as locomotives that run on tyres rather than rails.

— *Franklin Baldwin*

Baldwin

Baldwin tractors were in the big league, and for a time almost on their own. To start with, E.M. Baldwin and Sons Pty Ltd of Rooty Hill, north-west of Sydney, was the only tractor maker left in New South Wales in the 1990s. Then they custom built the most powerful locally made tractors in Australia.

The company started in 1928 as a general engineering business in some chicken sheds, on farmland near Castle Hill. They developed a creditable reputation in the field of sugarcane mechanisation, making cane locomotives of their own design. As the founder, E.M. Baldwin, grew older, activity gradually passed into the hands of his sons, Frank, Stan and Maurice.

The Baldwins were among the first in Australia to master techniques for stainless steel fabrication and they were awarded contracts for the manufacture of food and beverage containers. They eventually terminated one line, however, because it conflicted with the family's religious convictions. As Seventh-day Adventists, they were very concerned about the use of alcoholic beverages.

The original DM525, first Baldwin tractor for the agricultural market. This four-wheel-drive articulated tractor was powered by a Cummins engine of 525 HP. All-up weight of the tractor was just under 50 000 pounds (22 680 kg), unballasted.

The DP525 with same Cummins engine as the DM525 but Allison Powershift transmission instead of manual transmission.

In 1979 Frank Baldwin and the company's engineering team diversified into the agricultural tractor business with the launch of their first four-wheel-drive, the DM525. This articulated tractor was equipped with 10-speed mechanical transmission and powered by a Cummins KTA 1150 turbocharged and aftercooled 18.9 litre diesel engine of 525 HP. It was fitted with US-built Clark axles and final drives.

In 1982 the company received an Australian Design Award on an even larger model, the DP600, but they also produced a smaller model, the DM400. Both models received Australian Design Awards in 1984.

The DP600, the most powerful wheeled tractor in Australia at 600 engine HP, was Cummins KTA 1150 powered, driving through an Allison Powershift transmission. Smallest in the Baldwin stable was the DM300, a mere 290 HP at the engine. They built their own tractor final drive assemblies. The Baldwins were able to claim: 'To extend axle life, we incorporated cooling and filtration of oil in our axle drives. To the best of our knowledge no other manufacturer is doing this. It is expensive but cooling oil from a working temperature of 100°C to 65°C makes a lot

The DM400 equipped with Caterpillar 3406 diesel engine or Cummins NTA 855A, both rated 400 HP.

Baldwin's DM300 was the smallest in the Baldwin component-built tractor stable, at a 'mere' 290 HP from the Cummins LTA-10A engine, whereas the DP600 was the largest, with Cummins KTA 1150 at 600 HP.

of difference to the lifespan of your componentry.'

Despite some trading difficulties which followed the unprecedented slump in the agricultural tractor business of the 1980s, the company was able to produce mining and industrial equipment, and custom build tractors out of the factory, which was relocated at Rooty Hill in 1984. Tractor production ceased about 1990.

Phillips Acremaster

Laurie Phillips brought a tractor business to the remote town of Merredin, Western Australia. In 1975, Phillips produced his first Acremaster tractor there. It was the first of a series designed around Mercedes Benz power plants. By 1982 Phillips was offering four models of his big custom-built four-wheel-drive articulated tractors — the 402, 403, 404 and 404A.

Later in 1982 there was a shakeout in the company and a rival group was established in Merredin to make 4WD tractors. This breakaway group contracted to build their machines for IH Australia as International Acremaster tractors. By 1986 that series had expanded to include eight Acremaster models:

- C210M, Cummins 210 HP engine,
- M210M, Mercedes 213 HP,
- C250M, Cummins 250 HP,
- C275M, Cummins 275 HP,
- M275M, Mercedes 276 HP,
- M340M, Mercedes 350 HP,
- M400M, Mercedes 414 HP, and
- 520PS, Mercedes 513 HP.

With the takeover of International by Case in 1986, this tractor series ceased to be offered by Case-IH and Horwood Bagshaw took on the concept, offering in 1986/87 to component-build Acremaster tractors but only to tender.

Not to be outdone, Laurie Phillips re-emerged, from the ashes as it were, with a lineup of four new model tractors, appropriately named the Phoenix series. His new company was called Farmers Tractors Australia Pty Ltd.

In 1987 the articulated 4WD series were listed as follows:

- Phoenix 2100, Mercedes 210 HP engine,
- Phoenix 2800, Mercedes 275 HP,
- Phoenix 3500, Mercedes 343 HP, and
- Phoenix 4000, Mercedes 410 HP.

Phoenix tractors were only on offer for four years; now Phillips confines itself to tillage gear.

Phillips Acremaster 387, powered by a Mercedes Benz OM 400 engine in 1981.

[These big tractors can be seen in action in a fine video/CD series. See *Big Iron Part 3*, produced by DT-Media, Dwarsefeld, Germany. www.traktortotal.com.]

Zanetic-IH Acremaster, 1986.

Horwood Bagshaw Acremaster, front and side views.

26 RICHTERS OF BOONAH

Machinery 'built well to wear well'.
— *Company advertising*

Graham Richter.

Howard Richter.

Graham and Howard Richter carried on the family's custom-built machinery business in their hometown of Boonah, Queensland. The company built equipment to assist specialty-crop growers. Richters were proud of the fact that some of the original equipment — items such as corn, potato and onion machines made as far back as World War I or even before — were still in use half a century later.

Richter machinery was designed for integrated cropping systems, from planting, spraying, weeding, topping, harvesting, through to transporting and preparation for marketing: always with the aim of helping fruit, nut and vegetable producers to keep down costs on the crop of their specialty. Almost 90 different machines eventually found their way into the Richter catalogue.

Wherever possible, the Richter brothers designed their machines to use worldwide standardised components, such as drives, hydraulic fittings, bearings and wheels. Hydraulic drives were favoured for flexibility, and nowhere was this better displayed than in the Richter Hollagutz tractor and their TC22 Tool Carrier.

The Hollagutz — Nothing in the Middle

Complete farming systems could be developed around the Hollagutz as the power module. The chassis of this self-propelled tool carrier could literally split down the middle, opening out to enable the operator to drive unto or away from the attachment being used. Utilising the middle of the open-frame tractor had the advantage over rear- or front-mounted equipment of keeping machine length down and, being fully mounted, saved space and time on headlands. Power takeoffs were provided to front-, rear- or side-mounted equipment, as well as in the middle.

Three or more operations could be carried out simultaneously. The hydraulically split chassis design allowed the power module to be used in multiple applications.

The power plant of the standard Richter Hollagutz 60 was a 60 HP Lombardini air-cooled diesel engine, direct coupled to a Vickers transmission pump or

The Richters' Hollagutz or 'iron horse' tractor was hydrostatically propelled.

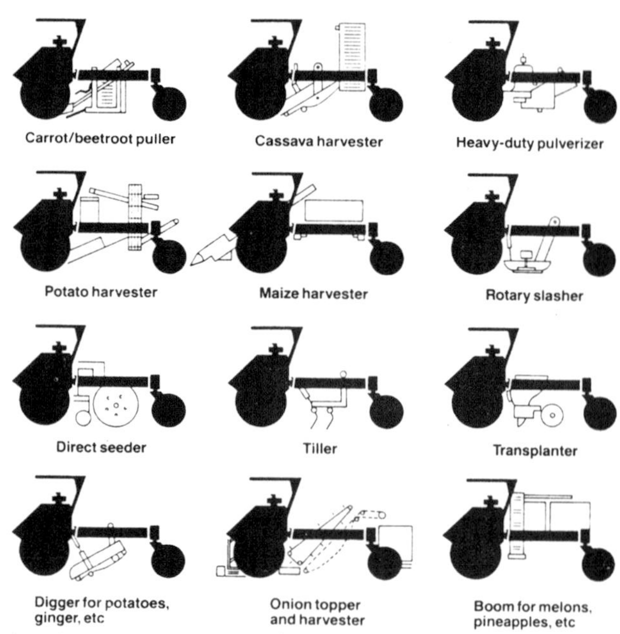

Some of the possible applications for the Hollagutz tractor. Richters offered the machine with two-wheel or four-wheel hydrostatic drive options.

pumps. Forward speed with the hydrostatic transmission was infinitely variable up to 9 mph, depending on wheel size. Optional four-wheel-drive articulated steering, or pivot steering with two-wheel-drive, was available. Another choice was engine power range, which could be supplied from 58 to 114 HP. Practically everything on the Hollagutz could have options.

The tractor could be fitted out as a mobile platform, or transporter, planter, weeder, sprayer or harvester.

Richters also custom built self-propelled tall-tree harvesting platforms, self-propelled cassava and bean harvesters, pull-type root crop harvesters, soil mixers, nursery equipment, shed processing plant, trailers, and custom-designed complete crop systems.

The Richter Brothers sold out in January, 1987.

Richter Hollagutz with picking and pruning boom, high-lift offset tipping hopper for palm, dates, nuts, coconuts, etc.

The Hollagutz tractor's unique and patented feature was the ability to drive away from or onto another machine.

Richter Palm-Fruit Tree Harvester.

27 AFTER-MARKET TRACTOR CABS — GASONS AND AUSCAB

In the biography of Frank Gason we see the story of an Australian worker who was grasped by an idea that would bring comfort and safety to men on the land.

— Rev. Dr Gordon Moyes, in the Foreword to Cab, Sir?, *1987*

Cab, Sir? — Gasons of Ararat

Frank Gason, managing director, A.F. Gason Pty Ltd.

A cab is not a tractor, so why a chapter on cabins? Nowadays a tractor can hardly be marketed without a cabin, and certainly not without a protective frame. Nowadays it is a legal requirement to have a roll-over protection structure on all new tractors above 560 kg. That was not the case when Frank Gason, motor trimmer and automotive body builder, established Australia's first tractor cabin works in Ararat. His first sale took place in 1947 when a Victorian farmer, Adam Stewart at Tatyoon, asked for a custom-built cabin for his Twin City model G.T.A. tractor. Gason sold him one with sliding windows, with the frame constructed from marine ply for durability against the elements.

The early Gason cabin design, c. 1951, with sliding windows. It was no oil painting, but it protected the operator from the elements.

Many old-time farmers regarded cabs as sissy, or an indication of weakness, so the cab business grew slowly, holding at about 60 a year through the 1950s. There was also some resistance by some who felt that the cabin structure would interfere with servicing the tractor.

The Gason cab design was first advertised in 1951 and changed little between that time and 1963. Nevertheless, at £49 (with the price raised to £54 after the first dozen were sold) those early Gason cabs were good value for protection against skin cancer. Australians have the dubious distinction of leading the world with this ailment.

Company director Frank Gason found that selling at Royal Shows was 'spectacularly unsuccessful' so he preferred the local farm machinery displays. There Frank could be seen promoting his cabins like a soapbox orator. He had more success with dealer outlets. The company switched to steel-framed cabs after 1961, gaining impetus from a contract to make rugged cabins for Caterpillar earthmoving equipment.

Frank's son, Les Gason, began his career as an apprentice with the firm in 1957, and became a business partner in 1962. That was the year they decided to specialise and concentrate on tractor cabin production. Their new steel designs for agricultural tractor cabs were launched at Wimmera Machinery Field Days in 1963. The five original employees were flat-out meeting the demand for 240 cabs that year. By 1968 the business had grown to a staff of 50, with an output of 2500 units a year. This was 62% of the tractor cabin market in Australia.

Introduction of ROPS

By the late 1960s over 100 farmers and operatives were being killed every year by tractors. An Australian Standards Association subcommittee, chaired by W.F. (Bill) Baillie, was established in 1969 to develop an Australian standard for roll-over protection structures (ROPS) for farm tractors. The former Tractor Test Station at Werribee became the official test centre for ROPS and three Gason prototype structures were submitted for test in 1971. All three failed, but valuable lessons were learned. With redesign, Gasons were able to claim to be the first to produce AS-1636 tested and approved tractor cabs in Australia.

Fortunes fluctuated for the company: 1968 was a

peak year due to the record wheat crop, but in August 1969 came the wheat quotas and farmers stopped buying equipment. Then in 1972 Deere launched their US-built cabins onto the Australian market. Their SoundGard cab posed a major threat for Gasons and the company sought tariff protection. But in 1974 Gasons were awarded a contract to build cabs for the Chamberlain C6100 and C670 tractors. Business picked up again and in 1976 Gason launched their own refrigerated air-conditioned cabin design, the Gason 30. Production peaked at 4000 units in 1978, and for this achievement the Victorian government gave them a Decentralised Industry Award. A special cabin was designed in 1977 for Ford Australia's unique hybrid tractor, the Model 8401.

The Filt-Aire option of the Econo-Cab could be purchased from Gasons to keep chemicals from the operator; it utilised an activated charcoal air-filtering system (below right).

Gason-produced OEM cabs for Massey Ferguson in 1978.

Gason's last tractor cabin design, the Econo-Cab, was state of the art, with optional air-conditioning, windscreen wiper and in-built ROPS.

Gason had 42 makes and models of cabin in its product range by 1980 through to 1986, but then cab production ended.

Diversification Into Tillage

In 1976 the Gasons decided to diversify by adding a tillage equipment line. They started to handle and build Hiniker tillage equipment. The move to diversify proved a sound business move, as tractor sales fell disastrously in the 1980s.

Meanwhile, sales of the scarifiers, harrows, mowers, and Aire-Stream planters made steady gains. The 1982 droughts, bushfires, rural recession and increased farmer dependence upon imports of machinery dealt another blow to the company and staff numbers fell considerably around 1983. The 1986 termination of the Chamberlain tractor line — the last of the locally built mass-produced tractors — was a particularly severe blow for Gason's cab business, even though they had won some business for fitting cabs locally on imported tractors, which were otherwise being brought into the country in roll-on/roll-off form, complete with cabs. Nonetheless, by 1988 an estimated 50,000 cabs had been sold by the company.

Frank Gason's story, so personally told in Schubert's book (1987), offers invaluable insights into facets of the tractor and farm equipment industry in Australia. Frank Gason's career has touched upon the lives of most of the luminaries on the tractor scene in this country. Frank's wish to operate with minimal business indebtedness may have tempered creativity to some extent, but it is a primary reason why his company has survived for over half a century.

There are many lessons there for Australian business.

Auscab

Graeme Logan started this business in Adelaide in 1992 as an offshoot from an air-conditioning business. It started with cabs for the 30-Series Kubota tractors that were coming in cab-less. Graeme Logan designed and made a ROPS-approved comfort cab that attracted a contract from Kubota. Nowadays the cab business has expanded to include cabs for road rollers, grape harvesters and many for ride-on tractors. Auscabs are tested to National Standards.

The ROPS Mandatory Requirement

Since the introduction of ROPS, the tractor fatality rate has halved in Australia. Briefly, after June 1996 all tractors used in a workplace must conform with AS 1636 (agricultural wheeled tractors rollover protection structures criterion and tests) and be fitted with a ROPS. Certain exemptions apply. Provided that the unladen tractor weighs more than 560 kg and less than 3860 kg and was purchased after 1 January 1981, ROPS is mandatory on the farm tractor regardless whether operated by an employee or not.

Examples (above and below) of the after-market cabs made by Auscab in Adelaide.

ROPS testing on the NATA-registered Gason test rig at the Ararat works.

28 SELF-PROPELLED AGRICULTURAL MACHINES

Not knowing what I was up against, I went ahead and did it.
— *R.G. Le Tourneau, 1960*

When Edward Clerk mobilised his portable steam engine he was probably the first to operate a self-propelled farm machine in Australia. Clerk had imported the Boydell-Garrett steam engine from England in the 1850s as a power unit to use on his 65,000 acre property at Bundarra. He reasoned that the massive Megaethon ought to be able to propel itself from the Port of Sydney to its destination, in northern New South Wales. He almost made it unaided, but oxen were required to assist for part of the protracted journey (Griffith, 1954).

Another decade was to elapse before imported steam traction engines were deployed on farms, but these were few, and for the most part used by contractors. Steam traction engines were large and costly. They needed a skilled and licensed operator, clean water and loads of bulky fuel. Many decades were to pass before Australian-built self-propelled (SP) machines were put to work in production agriculture, starting around 1913. As for the first self-propelled harvester, or combine, there is a common myth that it was of Australian origin. Frances Wheelhouse has added currency to the legend by writing that H.S. Taylor 'captured more honours by inventing the fully self-propelled auto-header in 1924 ... the idea of this fully self-propelled header was adopted by Americans in their machines' (*Digging Stick to Rotary Hoe*, 1966).

Unfortunately, this is far from the fact. Just who was the first with a purpose-built SP farm harvesting machine?

G.S. Berry and the World's First Self-Propelled Combine Harvester

George Stockton Berry — later Senator Berry, of Inigo, California — designed, built and used the world's first self-propelled grain harvester in 1886. Berry took five years to bring the concept to fruition on his 4000-acre 'bonanza' ranch in the Sacramento Valley. In the process he established more firsts than any other combine harvester builder before, or since.

Berry's steam combine, with 22-foot cut in 1886, was fuelled by the straw from the separator. In 1888 Berry wrote of this machine that it is 'as near perfection as a machine for that purpose can be made. The ground wheels are four feet face and six feet diameter. My header cuts a swath forty feet wide; it is made in two sections, and so arranged that it is handled with ease on very rough "hog wallow" lands, at the same time making a clean cut wherever it goes. The separator is of sufficient capacity to handle all the grain that can be got into it. I have averaged this season about ninety-two acres per day. I cut in two days 230 acres. It does not take any more men than I used last year to handle it, and it does about twice the work.'

His combine was:

- the first self-propelled combine harvester,
- the first straw-burning steam combine that was 'fuelled from the land',
- the first to employ a tractor to work both forwards and backwards; when he finished the harvest he used the tractor the 'right way round' for ploughing from the drawbar,
- the first tractor to use live power take-off, with a piped steam 'live drive',
- the first tractor with a differential in the drive axle,
- the first machine to harvest over 100 acres in a day,
- the first harvester to use artificial lighting to permit work after dark, and
- the largest header at the time: 22-foot cut initially, then 40-foot by 1888.

Berry had experimented with steam engines for motive power on other farm jobs before he built his first self-propelled combine. He deployed two engines on the machine — one of 25 HP on the traction unit and a 6 HP steam engine to power the harvester. Both engines piped their steam from the same boiler. This arrangement — a live power take-off — enabled the threshing machinery to operate steadily and independent of the tractor's forward movement.

Wood and coal were scarce in the valley, so Berry not only overcame the fuel problem but also saved himself considerable expense by designing the rig so that the threshed-out straw from the separator fell into a chute and was delivered onto a towed platform adjacent to the boiler, where it was stoked into the boiler grates manually. A supply of straw was deposited on the rack at night ready for the 30-minute task of getting up steam the following morning. As the rack filled during the day, excess straw was diverted and dumped onto the ground. Straw was also utilised as fuel when the tractor was used for tillage. For ploughing, the harvester components were removed and the tractor, now running forward, could easily pull a 20-bottom plough.

In 1888 Berry rebuilt the machine. He increased the cutting width from 22 feet to 40 and added headlamps for night work.

Local records indicate that four of Berry's combines were subsequently built by the Benicia Agricultural Works at a cost of $7000 each and were used by big grain growers until at least 1902.

When Berry turned his attentions to politics, he sold the rights of his design to San Leandro manufacturer Daniel Best, who utilised some of Berry's principles in his company's harvesters and steam traction outfits. In 1909, the first internal-combustion engined self-propelled combines were produced by the Holt Brothers in Stockton, California. In 1925, the rival Holt and Best interests were merged to form the Caterpillar enterprise. Caterpillar withdrew from the combine business 10 years later and sold their harvester interests and stock to Deere and Co., which added combines to their own product lines to eventually become the Western world's largest combine manufacturer (Quick and Buchele,1978).

Berry's influence was far-reaching. It is worth labouring the point about Berry's machine, to set to rest the myth that the first SP combine was developed in Australia. Local innovators were good, but unfortunately they weren't first. Nor was Massey Harris first — as is sometimes claimed in Canada.

Hugh Victor McKay built several self-propelled stripper-harvesters, but they were experimental and never produced in volume. The first(above left) appeared in 1909. Note tiller for steering. The same year Holt marketed their gasoline-powered SP in California. The second illustration shows a unit mounted on McKay's Model O tractor. The SP machine (right) brought the discharge from two stripper fronts together for delivery to the cleaning system.

Meanwhile Down Under ...

The man and the machine behind that self-propelled legend in Australia was Hugh Victor McKay, with his Sunshine Auto Header.

McKay first experimented with self-propelleds by re-working stripper-harvesters at the Sunshine works in 1909. He experimented for over a decade on several self-propelled harvester variants, but they were never put into production. One was evaluated as far away as Spain.

The Sun Auto Header

The Sun Auto Header was a joint H.S. Taylor–H.V. McKay development. These two innovators took out their first patent in July 1923, and others the following

The Sun Auto Header was introduced in 1924. The first models had a 12-foot cut and were powered by a four-cylinder Fordson engine. Ground drive was through a single wide driving wheel just behind the operator on the left hand side.

The Taylor-McKay patent on the Auto Header was issued in 1923.

year. Whereas the Berry combine had a side-feeding configuration, the Sun Auto Header was among the first commercial combines with a T-shaped configuration of platform and separator (there were similar designs being developed in Kansas by Curtis Baldwin and others a year or so earlier). The great virtue of this arrangement was that no crop was run down and there was symmetrical crop feed to the cylinder. The first production models, released in 1924, had a 12-foot cutting head and were propelled by a Fordson engine. Ground drive was through the single driving wheel behind the operator. The Sun Auto Header continued H.S. Taylor's comb-front and twin-auger feed combination that had first attracted McKay to hire the Henty-born inventor to work at Sunshine and to bring that comb front design with him. Production of the Auto Header was discontinued right at the end of World War II, after 1356 units had been built.

The Squire Autoplough: A Self-propelled Powered Disc Implement

Quirindi (NSW) farmer R.A. Squire is credited, at least by the Australian press, with having invented the combine drill. Squire was intrigued by possibilities for combining cultivation and seeding and in 1913 he patented a 'combined tractor and agricultural implement'. This self-propelled machine consisted of an engine-powered disc over the top of which was mounted a seed box and grain metering system to drop the seed through tubes adjacent to the discs. His first field-going prototype was built in 1917. Tests and machine improvements went on until at least 1929 or 1930.

In 1930, *The Australian* newspaper cited Professor Watt of Sydney University, who credited Squire's Autoplough with 'revolutionising the agricultural industry' (26 April). Evidently the professor was unaware of the Hackney Auto-Plow, the Birrell, and other US-built tractor-ploughs of similar configuration to Squire's — and they went as far back as 1911.

Squire used a secondhand Buick motorcar engine of 27 HP to power the front driving axle through a car gearbox and chain-drive system. The midmounted disc gang was powered by driveshaft, bevel gears and lay shafts. The nine 26-inch diameter concave discs turned in the same direction as the tractor drive wheels, but 16% faster. The disc-gang and its drive were mounted on a hinged frame, held down by springs to allow the disc-gang to 'jump rocks or stumps' and absorb shock loads.

The plough could work loamy fallow at between 2 and 4 mph, pulverising the soil to a depth of eight inches. The discs could be left in the ground during turns without being disengaged. At the rear of the machine was a single steering and depth-control wheel which ran in the furrows.

Around 1929 Squire had established the Squire Company, with a business address at 16 Barrack Street, Sydney, but it became a victim of the Great Depression. There is no indication whether any machine was built other than the one exhibited at the Sydney Royal Easter Show and tested at Hawkesbury Agricultural College. The inventor had plans for an improved drive system, and intended to utilise the concept for seeding, fertilising

The Squire Autoplough, as demonstrated at Hawkesbury Agricultural College (HAC), Richmond, NSW, in October 1929 before a gathering of the Power Farming Association. Squire's machine had a powered disc-gang rotating in the same direction as the front drive axle, but 16% faster for good soil pulverisation. (Photos courtesy HAC)

The 1911 Oppsata Motor Plow. Two models, of 15 and 30 HP, were offered in 1911 by this Wisconsin manufacturer of three-wheeled midmounted auto-ploughs. This was just one of several US makers of ingenious self-propelled implements in the 1911–12 period (Gray, 1975). These machines anticipated Australian designs by a few years and were influential on local inventors.

and harrowing in a one-pass operation (*The Australian*, 26 April 1930).

Rein-drive Tractors

The rein-drive tractor concept was a sidetrack that diverted many erstwhile inventors who were seeking to replace the horse with horsepower. In one popular form this was an arrangement using a single-axle, two-wheel-drive tractor hitched ahead of the implement. The operator rode on the implement and controlled the tractor by reins or ropes hitched to control levers. Driving such an outfit was quite an act, especially when the reins were also utilised to change gears!

An early US example of a rein-drive tractor was the Hoke of 1913 (Gray, 1975); others were the 1913 Cole, the 1914 Detroit and the 1915 Line Drive. The 1918 Rein Drive from Canada also followed this principle and was a machine which Mallee machinery collector A.M. Gregory reports was at one stage considered for manufacture in Australia under license by Kelly & Lewis of Victoria.

Cornelius Murnane of Haberfield, near Sydney, sought a patent in 1914 and another in 1915 for 'Improvements in tractors for agricultural implements and the like.' Murnane's design was a rein-drive tractor. The two reins controlled forward motion, braking and steering. Kelly & Lewis reportedly built an engine for Murnane's use.

What makes Murnane's effort noteworthy is that in 1924 he managed to persuade the venerable British firm of John Fowler of Leeds to build his design under licence. Two pre-production pilot models were built by Fowlers. One was exhibited at the 1924 Royal Agricultural Society Show at Leicester, where it won a silver medal in the 'Most Improved Farm Machinery' section. The Fowler Rein-Drive with its 32 HP engine was reputed to have pulled a three-furrow plough, and could be hitched up successively to a cultivator, seed drill, reaper, baler or fully loaded wagon.

Two machines were shipped to Australia around 1926, where they were to be marketed by Austral Engineering Co. of Kent Street, Sydney. The units could be purchased for £650, including pto drive and a two-ton capacity trailer. One of the Fowler Rein-Drives survives in the Pioneer Park machinery collection at Parkes, NSW.

Another rein-drive tractor was developed in Australia along the lines of the 1918 Moline tractor. It was home built by the Moline tractor dealer in Horsham, J.F. O'Callaghan. In 1927, a tractor trial was conducted at Longerenong Agricultural College where O'Callaghan's 30 HP diesel-powered unit was said to have displayed 'a good reserve of power' pulling two four-furrow ploughs coupled together and working at four inches depth in stiff black soil. The cost of fuel was estimated at 8 pence an acre. Ploughs and tractor were reportedly 'easily handled by one man'.

The single-axle, two-wheel tractor concept was only a diversion in Australian broadacre farming. Its principal

*The Fowler Rein-Drive tractor was said by the manufacturers 'to couple up easily and quickly to all existing types of machines or implement'. Fowler's Rein-Drive was powered by a Fowler V-type twin-cylinder engine of 32 HP at 1000 rpm. It was developed initially by Australian Cornelius Murnane, whose machine is seen in the lower illustration (*Power Farming*. 25 November 1925).*

advantage was that it enabled one-man operation of tractor and implement, but in Asia this tractor concept is used widely and there are literally many millions of single-axle tractors in use.

Lennon's ABC Auto Reaper-Thresher

Hugh Lennon Pty Ltd, Spotswood (founded about 1870),· was a farm machinery maker of modest profile which offered most of its production through International Harv-ester Australia. But in 1930 they launched a self-propelled combine under their own name, developed from a farmer's design. The header was mounted over a Fordson tractor. The front axle cluster was removed and attached to the rear of the reaper-thresher. Cutting width was 14 feet, which gave the machine a field capacity of 50 acres a day. But the Lennon-Stapledon ABC harvester enjoyed only the briefest share of the limelight. The company folded in 1935, a casualty of the Great Depression.

Another short-lived machine of similar style and vintage was the Murnane and Webb Tractor-Machine. The Melbourne-based inventors and designers of the machine put their first model on display at the Melbourne Show in 1931, on the Cliff & Bunting stand. It was touted as 'the perfected grain grower's complete power farming plant, convertible into a tractor-harvester, tractor-cultivating drill, tractor-plough, tractor-scarifier, power wagon, timber jinker and other combinations to suit seasonal work'.

The Lennon-Stapledon ABC Auto Reaper-Thresher, c. 1930. Company advertising said of the ABC that it 'cannot raise the dead but it can lift the fallen', referring to the machine's capacity to lift lodged crops. Both the bulk bin version and bagging model were powered by a Fordson tractor.

Connor-Shea's Auto Header

There were major upheavals in the 1950s at the Sunshine Harvester works, when Massey Ferguson took over from the McKay management. Two engineers, Tom Connor and Les Shea, left the company to branch out on their own. They established Connor-Shea Limited at Sunshine in 1952 to build a range of cultivating, seeding and harvesting equipment — and a tractor-mounted harvester. The tractor was reversed and the header was mounted over the top, with the controls and operator's

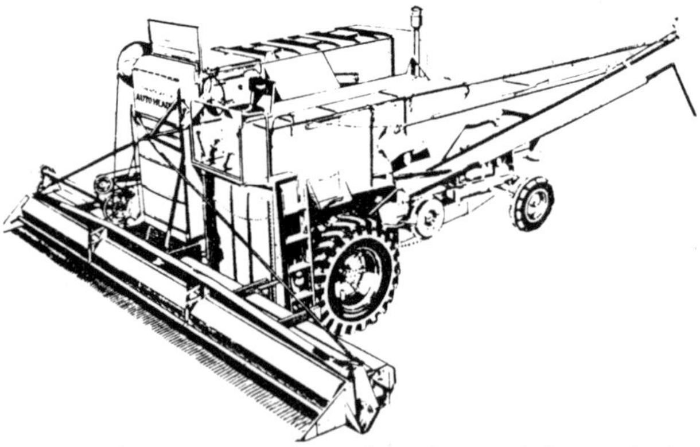

Connor-Shea's Auto Header with 27 foot comb front, which was a record width for the time (ca. 1970), although the 23 foot front was the more popular size.

Jondaryan Woolshed possesses the only surviving Road-Burner. A series of burner nozzles faced directly down on the road surface to cook the clay and render it hard enough to allow traffic to get through in the wet part of the year on what would otherwise be impassable tracks. It was in effect a mobile kiln. Much of the lower structure comprised fire-brick shields to concentrate and prevent heat from escaping out the sides and from scalding the top. The 37-ton machine was 35 feet long, had a 23-man crew and was powered by two engines.

station strategically located as on a regular combine harvester.

Although the Connor-Shea Auto Header was discontinued in 1976 and production never exceeded 50 units per year, it was a significant contribution to the technology. It had the following features: widest gathering front on a postwar combine (closed fronts from 15 to 27 feet wide were available); the gathering front and elevator were raised parallel, up to a maximum transport clearance height of 53 inches on one version, in order to clear fenceposts. It was possibly the fastest combine in its day, being able to utilise the tractor transmission with road speeds up to 20 mph. It was also designed with ease of cleanout in view and could be cleaned out faster (four hours) than any other combine harvester at the time. This cleanout feature was and still is an important feature in Australia, where insects over-wintering in combine residues may re-infest grain harvested at the beginning of the next season, to the detriment of our grain export markets.

A Weird One-off at Jondaryan, Queensland

The Jondaryan Woolshed and Museum, on the rich and productive black clay-loam soils of the Darling Downs, is the final resting place for a strange self-propelling machine. This monster is a Road-Burner. It was designed to travel down a clay road and literally bake the clay to make the road trafficable in the wet seasons. Two of these Road-Burners were built in Sydney in 1927, and they are said to be the only ones in the world. Only one survives. The 37-ton monsters were propelled by a 16 HP petrol engine at the colossal working speed of 24 feet an hour — which comes out at 220 hours per mile! Another engine (18 HP diesel) powered the blast fan and pumps, etc. A round-the-clock crew of 23 accompanied the machine, which baked the ground surface at up to 2600^{0}F. One road so treated became known as the Red Road after it had been surfaced by Sydney inventor Irvine's Travelling Furnace or Road-Burner.

CTF and Gantry Tractors

Wheel tracks in crops cause compaction and can lower yields on vulnerable soils. In Australian agriculture today a hot topic is controlled traffic farming (CTF) — that is, a system of crop farming in which equipment wheel tracks are confined to 'roadbeds' down the paddock. The cropped area in between, the 'rootbed', is not driven over or compacted by tractor wheels. The wider apart the tracks, the less is the relative proportion of unproductive land taken up by the roadbeds. The advantages of such a cropping pattern are:

- traffic-induced compaction is eliminated from the cropped area or rootbed,
- less fuel and time is required for tillage, because draft is reduced by as much a 40%, and there may be less operations,
- greater cropping flexibility,
- there is less tyre rolling resistance when operating an a set path, and varying weight of the machine does not alter depth control,
- field access (trafficability) is improved and therefore timeliness,
- ability to enter the crop up to full maturity if need be,
- there can be improved yield due to better soil texture, increased water infiltration and moisture retention,
- improved implement control and stability, and
- potential to reduce implement cost and use lower powered tractors.

A gantry tractor on the other hand is a purpose-built self-propelling wide-span tool frame, usually with wheels that swivel for end-on road travel.

Alfarm-Mizzi self-propelled sugarcane harvester for green cane harvesting. Only a handful were ever sold.

This cane harvester stamp was issued in 1991. Very few Australian stamps have been issued showing tractors or farm machinery, maybe three in total, unlike some countries that have had essentially no farm equipment industry, like Mongolia and New Zealand.

Goldacres Trading Pty Ltd in St Arnaud, Victoria, are manufacturers of trailed spraying equipment and self-propelled boom sprayers. Goldacres make four SP models. The 4WD SP model shown has a boom 33 feet wide that is air-suspended.

The Dingo mini-loader is made by Dingo Mini-Diggers Pty Ltd in Dalby, Queensland, and exported widely. Model K9-3 is shown on the front cover. Here is a purpose-built remote-controlled model K9-4 for situations where conditions are too hazardous for the operator's direct presence. Launched first in Australia, the market for mini-loaders has expanded enormously and now there are at least 12 makers involved in the United States, as well as the two Australian manufacturer/licensees, Dingo and Kanga, both based in Queensland.

Three-metre wheel tracks. Jimbour farmer Jamie Grant's JCB Fastrac tractor. It is not Australian built but adapted to specific conditions by widening the wheel track to 3 metres to match the harvester and reduce soil compaction in the controlled traffic farming (CTF) system that Grant uses for his total cropping program on the Darling Downs.

A John Deere 8520 tractor with axle spools to extend the track width to 3 metres for controlled traffic farming (CTF) on Rob Taylor's property on the Downs.

The concept attracted a small amount of commercial and academic attention as innovators built gantry tractors that could span the rootbed with a bridge-like tractor frame which carried the implements, sprayers or crop processing equipment. There were several Australian gantry tractor designs. One self-propelled tomato harvesting platform was 15 metres wide, another 13.5 metres. These machines travelled end-wise for transport, but they have not been taken up in quantity by manufacturers. It is salutary to be reminded of Englishman Henry Grafton, whose 15-metre steam-powered gantry concept was tested in 1860.

Neville Toft's SP Picker Platforms

Neville Toft's business was an independent offshoot from the long-established Toft clan of Bundaberg, Queensland. His company specialised in self-propelled picking aids for fruit and vegetable crops. With these machines in their area, stoop labour became past history.

An asking price of $185 000 might have seemed a lot for a self-propelled skeleton of a tractor, but closer scrutiny reveals complex hydrostatics, ingenuity and good value for a tomato-harvesting machine that could pay for itself in a season. Neville Toft's initial prime mover was a 50 HP Deutz powerpack driving hydrostatically to all four 23.1 x 26 wheels. This model could carry 18 seated pickers on nine trailing booms. Each boom can be lowered independently. The produce was gently delivered by conveyor belts to the bulk bin. A sorting section enabled reject fruit to be dispatched before the bulk bins reached the packing shed.

Toft also built lighter duty three-wheeled versions, with 50 HP power pack (again fully hydrostatic) and carrying 10 pickers, capable of operating all day on four gallons of distillate. Both designs were fully articulated

The four-wheel-drive Toft platform tomato harvester with nine booms to carry 18 pickers.

Picking platforms raised for clearance or transport. The platform swivelled on a turntable for road movement.

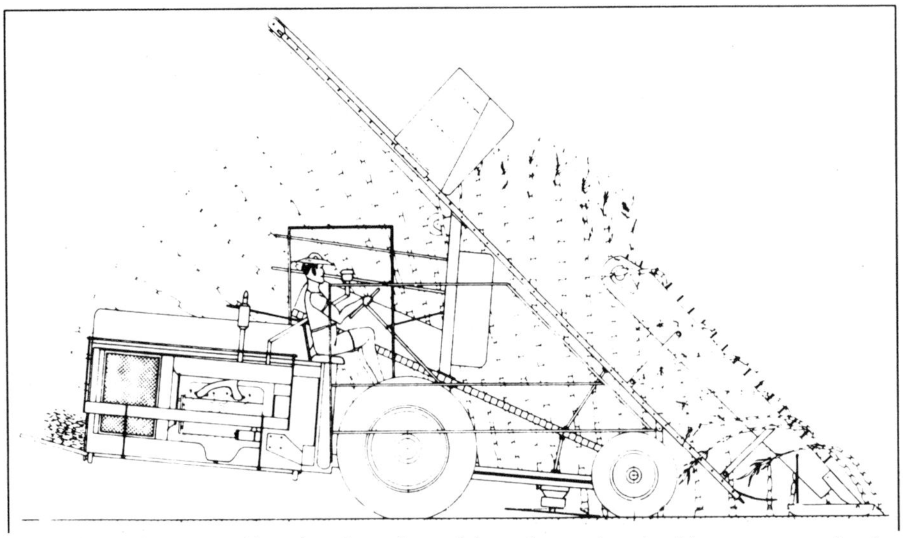

The Toft Brothers, Harold and Colin, of Bundaberg designed and sold an ingenious loader in 1956. It was mounted on a Blitz truck. Toft went on from there to manufacture self-propelled harvesters in 1960. Their J.150 wholestick harvester with 40 HP International power unit is illustrated. During the 1960s Bundaberg became a world centre for cane harvester production. The Australian sugarcane crop was soon practically all mechanically harvested. Hand cutting became a novelty within a decade.

and could turn on their lengths. In transport mode the harvesting platform rotated on a turntable which brought overall width below 12 feet so that they could be driven on public roads without police escort. Neville Toft Engineering also made self-propelled sprayers and single operator fruit pickers (cherry-picker type). The last vestige of the Toft-type SP harvester platforms is to be found in Kyabram, Victoria, where JN&R Engineering makes such machines.

The story of the mechanisation of the cane harvest through to 1985 has been assembled and thoroughly told in *The Canecutters* by Geoff Burrows and Clive Morton (1986). The early stages and very recent cane harvester developments will be discussed in the next chapter, insofar as they relate to Australian designers of agricultural prime movers.

JN&R

JN&R Engineering, of Kyabram, Victoria, have been making self-propelled tomato harvesters (platform types) since 1992.

Summary

Australia has an intriguing heritage of farm equipment developers, as is indicated by this sampling of self-propelled machines. Of all who have tried, few have survived as viable enterprises. Just a handful of local SP manufacturers are left.

Were those developers who didn't make it naive? Perhaps. Many of these locally built products were built by entrepreneurs who, like R.G. Le Tourneau, did not know what they were up against, 'so went ahead and did it'. Some have argued that more government intervention may have helped. Whatever the case, corporate rationalisation has made Australian agriculture heavily dependent on overseas developments for new equipment.

In today's broadacre machinery market, tillage, compact tractors and self-propelleds may be the last bastions for Aussie ingenuity in farm machinery.

JN&R Harvester — a seven-chute tomato picker platform, in transport mode.

JN&R self-propelled row-crop trellis-pruner.

29 AUSTRALIAN SP CANE HARVESTERS

Buy Australian goods.
— patriotic epigram cast into seats of some early McKay machines

Mechanising the Cane Harvest

Australian inventors were at the fore in the mechanisation of the sugarcane harvest. Abundant details can be found in articles by Clive Morton in 1979 and by Ken Blyth in 1983–84 in a series in *The Australian Canegrower*. There is a also fine monograph by Burrows and Morton, *The Canecutters* (1986).

Sugarcane was harvested by hand when first produced commercially in Australia in the 1860s. For the next 90 years indentured labourers wielded wickedly sharp cane knives. The field was usually hot-burnt first to reduce leafy material, enhance sugar content and get rid of venomous snakes and rats. The juicy stalks needed to be transported rapidly to the mill for juice extraction to reduce losses out of the cut stalks and minimise organic decay. Morton (1979) relates how the canefield mechanisation story began in 1899, but he makes the point that 'the technology of the times did not match the dreams of the designers'.

A key player on the canefields was Ralph Sadlier Falkiner, of Boonoke merino grazing property fame. Falkiner must have had a passionate interest in farm

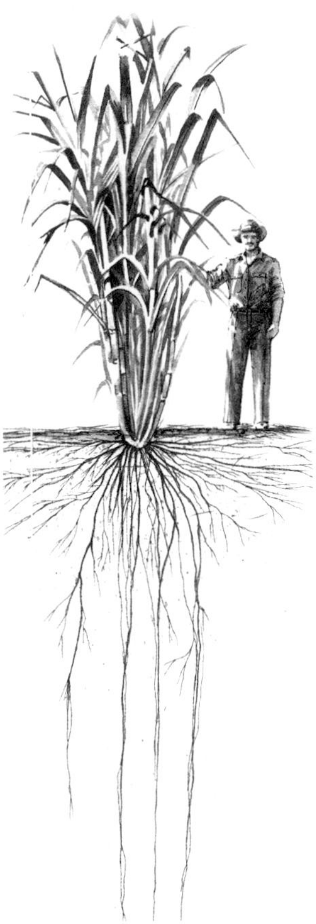

Sugarcane plant. (Queensland Sugar Corporation)

Manually harvested sugarcane at the mill with Kanaka cutters, early 1900s. (Australian Canegrower)

One of R.S. Falkiner's self-propelled sugarcane harvester prototypes, c. 1927. This machine pioneered mechanised chopped cane production. Falkiner did not succeed in commercialising his machine in Australia, where the first truly commercial self-propelled machines did not appear on canefields until around 1960. (Australian Canegrower)

machinery. He supported the Jelbarts with their tractor developments through difficult years in Ballarat, imported an enormous electric-wheel road train, and assisted the tenacious Frank Bottrill with his massive Big Lizzie tractor with its Dreadnaught wheels. In 1921, Falkiner took over the struggling Hurrey sugarcane harvester project in Queensland. Field trials of the first whole-stalk machines, and then chopper-harvesters, proceeded through to 1929. After building six units, Falkiner elected to concentrate on overseas projects to promote the technology. He did so because he felt there was insufficient interest in Australia. Falkiner had cane farms at Clewiston, Florida, where he had been carrying on concurrent field trials of the harvester. He ordered 20 machines for his North American activities in 1931 and the Allis-Chalmers Company set about building those under licence.

Howard's Austral whole-stick cane harvester.

Howard's Austral Cane Harvester

Arthur Cliff Howard, the designer of the world-famous rotary hoe that still bears his name, built many machines apart from Rotavators — engines, tractors, washing machines … and cane harvesters.

Queensland's canefields were a reliable market for his rotary hoes and tractors, so it should come as no great surprise that the energetic inventor would turn his hand to mechanising the cane harvest. Howard's first cane harvester was demonstrated in November 1929 at Bundaberg. But several prototypes had been fabricated at the Northmead works for testing in Queensland before that first public demonstration.

Howard was encouraged in his developmental program by the Fairymead Sugar Company, which provided financial support (as they also did for Falkiner) to foster development of mechanical cane harvesting. Howard's machine, unlike Falkiner's, was extremely compact, but then it was a whole-stick machine and not a chopper-harvester.

Undoubtedly, the efforts of pioneers such as Howard and Falkiner made a significant contribution to the mechanisation of the cane harvest. But it took until around 1960 for a truly commercial machine to reach the market. Falkiner's work was the springboard for the Creber Bros of Mackay; for International Harvester engineers and for Ken Gaunt, who subsequently spearheaded the Massey Ferguson team's efforts on their chopper-harvester. Don Quick (the author's brother) was a key engineer in the developments, which led the cane harvester industry in their time. Hand cutting became a novelty in the 1970s. By then cane was being delivered to the mills from the chopper-harvesters as billets 250 to 350 mm long. Unfortunately, after selling some 1500 SP machines Massey Ferguson pulled out completely in 1984 and left it to Austoft for serial production of cane harvesters. Their output peaked at around 160 units a year. Austoft was subsequently taken over by Case International. So Bundaberg's years as the world centre of cane harvester engineering and production for the global market were of limited span.

Alfarm-Mizzi self-propelled sugarcane harvester for either green or burnt cane. Only a handful were ever sold.

Massey Ferguson's prototype ZX-11 green cane harvester and MF Model 105, both wheeled machines. (Courtesy Don Quick)

The Alfarm-Mizzi Universal Cane Harvester

Ingham canegrower and inventor Joe Mizzi developed a self-propelled cane harvester to harvest both burnt and green cane under varying ground conditions. Alfarm of Albury sought to diversify their product lineup and undertook this project in 1987 as a product which they perceived had potential for export. Power unit was a 320 HP Detroit-Allis V6 diesel powering an extensive group of hydrostatic drive units for ground drive (through the rear axle), crop gatherers, base cutters, toppers and trash extractors.

The Rise and Fall of Case-Austoft

The Australian sugarcane crop was 100% mechanically harvested by 1979 (Queensland Sugar Corporation) and much of that equipment development came out of the Bundaberg district. Toft's formed Austoft and faced many fiscal twists and turns, but its local and export sales were enough to attract a buy-out by Case-IH to set up as Case-Austoft in 1997. Then in 2004, Case-Austoft withdrew production of cane harvesters from Bundaberg to move production to Brazil. The world decline in sugar

325 HP Komatsu-powered Case-Austoft 2000 model harvester on tracked undercarriage for handling wet conditions in the high rainfall canegrowing areas.

Case-Austoft SP Powerhaul wheeled haulout, 170 HP Cummins powered.

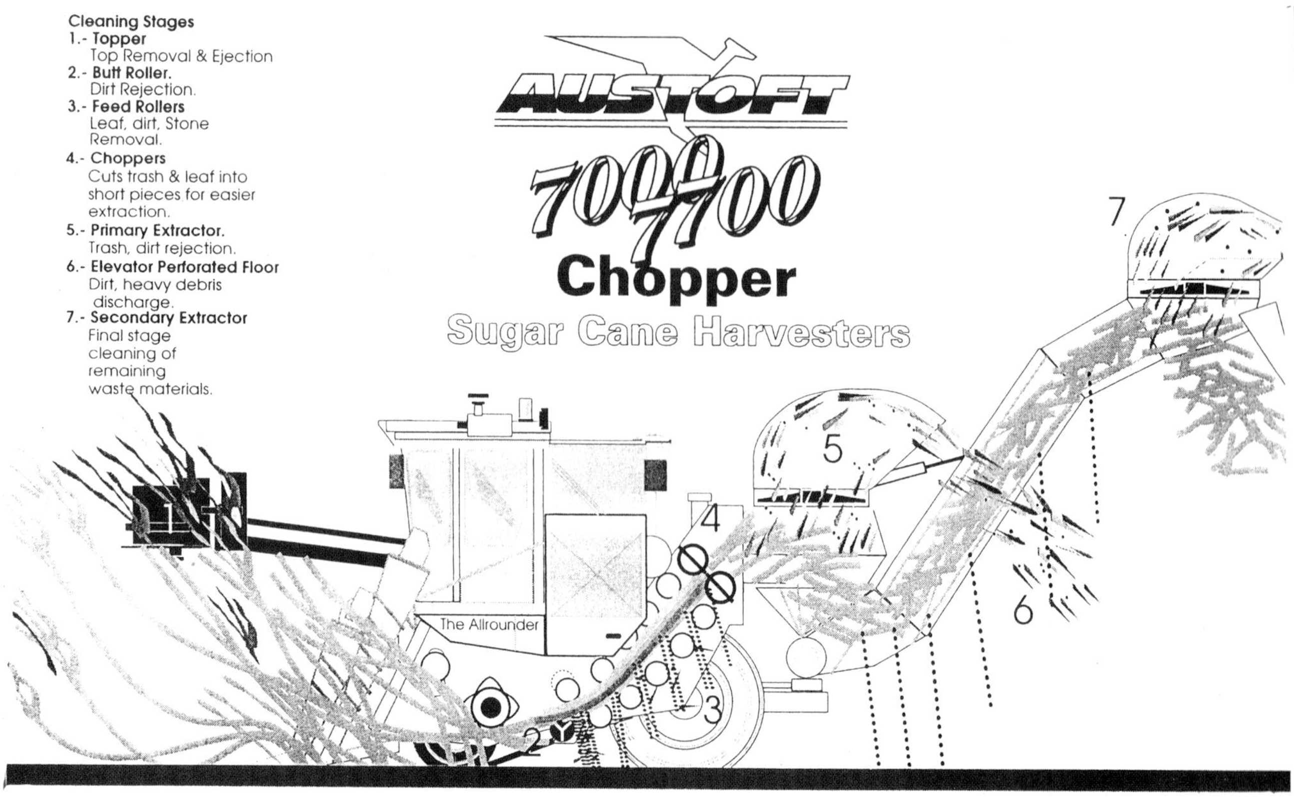

Case-Austoft single-row cane harvester schematic.

prices around the turn of the millennium and the shifts in the main market for harvesters were key factors.

Self-propelled Haulouts

A feature of Australian cane harvesting has been the extensive use of self-propelled haulouts or cane bins (as well as trailed versions), whereas in the US and elsewhere there is more reliance on tractor-drawn/trailed bins. Case-Austoft had a handy market for SP wheeled and crawler-tracked bins. Carta's at Ingham make an ingenious trailed cane haul-out.

Doug Young's Two-row Cane Harvesters and Transporters

Bundaberg canegrower Doug Young and family have built from scratch or remodelled a number of cane harvesters and cane transporters over the past two decades, mostly for their own 350 acres and for their cane harvest contracting business. The latest harvesters are two-row models, as distinct from the industry standard single-row machines.

Doug Young's self-propelled two-row machines have refined principles over the single-row, and have greater handling and cleaning capacity. The main cleaning fan, for example, is 6 feet 6 inches in diameter. Power comes

Bundaberg canegrower Doug Young's purpose-built cane harvester and SP transporter work together at harvesting speeds of around 6 mph. It takes two transporters full-time to keep up with the harvester.

Doug Young stands beside the towering 530 HP two-row cane harvester that was built in his workshop. He has made several harvesters and transporters in the past decade. The 250 HP self-propelled haulout on the left is another product of his ingenuity.

from a modified 17 litre Iveco engine boosted to around 530 HP. He has built three SP side-unload transporters, each powered by Iveco 200 HP engines. One harvest setup like his would cost around $1.2 million, or to contract, a custom rate of $870 an hour. Travelling at 6 mph the cane can come off at rates up to 200 tons an hour.

Cameco in Louisiana, a subsidiary of Deere, has acquired the rights and used Doug Young's expertise to build 10 market-test versions of these two-row harvesters in 2002/03.

Summary

Sugar is Australia's second largest export crop, after wheat. Ninety-five percent of the raw sugar output is grown in Queensland and is worth around $2 billion annually. Australia is one of the world's largest exporters of raw sugar.

Sugarcane is a grass that is highly efficient at trapping the sun's energy and converting it into sucrose, more so than most other crops. The plant needs lots of sun, a frost-free climate and at least 60 inches (say 1500 mm) of water as rain or irrigation a year. The cane plant stores the sugar in its fibrous stalk. Yields are typically 80–100 tonnes per hectare, from which 10–15 tonnes of sugar are produced (Queensland Sugar Corporation). To give a comparison, a top-yielding corn crop may produce 15 tonnes of fibre, while a good field of alfalfa hay may produce 20 tonnes of biomass from one hectare. It will be therefore be appreciated that the amount of material that a cane harvester has to handle is prodigious. At first the cane was harvested by hand with the labourers wielding wickedly sharp cane knives. The field was usually hot-burnt first to reduce leafy material, enhance sugar content and get rid of venomous snakes and rats. Nowadays all Australian sugar is mechanically harvested by self-propelled harvesters and sophisticated transporters. Conversion to mechanical harvesting was complete by 1979.

A downside to Australian sugarcane farming is its vulnerability to world sugar prices which can fluctuate wildly, for example, 6.5 US cents per pound in 1999 versus 13.43 cents in 1995. Sugarcane production is hugely affected by competition from European sugar beet and by consumer dietary trends.

There was a time when Bundaberg was the world's centre for sugarcane harvesting equipment. Massey Ferguson, Falkiner, Howard, Mizzi, Toft, Case, Young, Carta ... are names that have flashed across the harvest screen of the Australian canefields. After selling several thousand SP units, unfortunately no local builders remain in the cane harvester production business.

30 SPECIAL CROP SPs: OLIVE AND GRAPE HARVESTERS, AND BROADACRE SPRAYERS

The supply of tractors and all types of agricultural machines is far behind demand.
— Power Farming, *February 1949*

Olives: Task Machinery

Tony Hawke and his team at Task Machinery at Wauchope on the mid North Coast of New South Wales developed the Hawke Olive Harvester between 1999 and 2002. The Hawke Harvester is designed for mature large olive trees. The self-propelled machine encircles the tree and the oscillating fingers work their way up, removing the fruit. One of the major advantages of the Hawke Harvester is that it allows olive trees to grow quite large — for example, trees up to six metres tall and nearly five metres wide. This gives growers the ability to maximise fruit yields on a per hectare basis, for more profit from the grove. The harvested yield — typically 30 kg per six-year-old tree — is superior to traditional methods and there is essentially no damage to the tree.

As the shaker ring makes its way up the tree, 90 rods are hydraulically actuated to shake the branches 'sinuously'. Catching frames gently collect the olives and convey them into the bins. Capacity is up to 150 trees an hour. The machine costs around $300 000 in self-propelled 4WD configuration.

In Australia and California, there was no way that olives could ever become a significant crop unless the harvest could be fully mechanised. Reliable hand-harvest labour is simply not available for the larger orchards.

Mechanised Grape Harvest

An estimated 100 new grape harvesters are sold in Australia every year, with imported Gregoire models commanding the major share. There may be about 1000 grape harvesters in vineyards, and the market is now at replacement-only level. One reporter claims the market is oversubscribed. However there is one local maker.

UR Machinery Pty Ltd, of Mildura, northern Victoria, have been making pull-type grape harvesters since 1992 and recently merged with Kverneland Group, which started in Norway a century ago and now claims to be the world's largest maker of specialist farm machinery.

he Hawke Harvester built by Task Machinery, Wauchope, NSW, is designed for mature large olive trees. It can harvest up to 150 trees an hour. The harvester uses 90 shaking rods to dislodge the olives into the split catching frame.

They offer both the Gregoire and the Australian-made Vinestar grape harvesters. The Vinestar 800 is a tractor-drawn single-row harvester. The Vinestar 1000 series is the self-propelled model made in Mildura and starting to be exported to several countries.

Vinestar 1000 is a new compact design to meet local transport regulations. It has an air-conditioned cabin and novel steering geometry to improve manoeuvrability. The 1000 series has an 800 mm (32-inch) through-passage picking head, suspended conveyor system and stepped bucket-type cross conveyor with cleaning fan.

Goldacres' self-propelled boom sprayers have five tank size options, from 2000 to 4000 litre capacity, and booms from 24 to 33 metres wide. The drive to the four-wheel ground drive is mechanical, while the boom is air bag suspended.

The picking head can be taken out quickly and be replaced with pre-pruners, sprayers, hedger, mulcher and under-vine cultivators to make the prime mover a year-round workhorse in the vineyard. A soft pick minimises trellis damage.

Self-Propelled Field Sprayers

Australia has just one local maker of SP spray rigs: Goldacres Trading Pty Ltd. The St Arnaud, Victoria, agricultural equipment manufacturer has been making trailed sprayers since 1993 at their St Arnaud and Ballarat plants. In 1998 Roger Richards started on the design of an SP model. By 2000 the prototype was displayed at the Wimmera Machinery Field Days. Orders came in rapidly and now the company has four models with a range of options. By 2005 Goldacres had made 80 SP sprayers and the order books are full.

Their product line leaders, trailed and SP, have boom widths from 18 to 33 metres, tank sizes 2000 to 4000 litres and are offered with a range of equipment choices. The most advanced is their 3TS system which automatically matches spray pressure to forward speed for spray application uniformity. Factory upgrades came about with the help of the Victorian government's Grow Your Business programs for manufacturers in regional Victoria. This enabled Goldacres to achieve a 20% increase in sprayer manufacturing levels. Their SP sprayer can also be fitted with hands-free Auto-Steer by Beeline, another Australian development.

View from the operator's perspective from the Goldacres SP Sprayer. Sophisticated controllers in the cab govern and indicate boom performance. The Goldacres 3TS dual-line system enables three flowrates progressively, in accordance with forward speed, to maintain correct droplet size and application efficacy.

Summary

This sampling of self-propelled machines does not include all the local machines that have appeared over time. It deals with the ones that remain in 2005 and for which information was available.

It has to be noted, however, that the total numbers of these SP machines is very small compared with the actual numbers coming into the country. But for those rugged local SP manufacturers that are left — and their employees — this is a vital business. With the technologies changing so fast, it is a great credit to them that they match and even surpass the competition.

31 COMPACT TRACTORS

The bright spot is the outdoor power equipment (OPE) and compact equipment business, as the number of small acreage properties increases and demand for compact industrial equipment grows.

Compact equipment includes small or utility tractors, ride-on mowers, skid-steer loaders, ATVs, hand-held or pedestrian machines, and so on. And here Australian makers do have a significant piece of the action.

Growing numbers of people want to get out of the cities to live a country lifestyle. That, and urban gardening, have fuelled demand for compact equipment. There was a growth spurt for compact machines in Australia in the 1990s and, while demand has fluctuated with the seasons, ride-on mowers sales are in the range of 30–40,000 units a year. Those numbers are not included in the Tractor and Machinery Association's tractor statistics, where total tractor sales struggle to reach 10,000 a year. As for pedestrian-controlled mowers (backyard engine-powered walk-behind, or push-mowers), the demand is for almost as many mowers as there are new car registrations in Australia — at more than 350 000 a year. Then again there is a healthy market for all-terrain vehicles (ATVs), all 14 000 of which are imports. Finally, there are mini-skid loaders — a market dominated by local makers — not to mention locally made portable lawn edgers, leaf blowers and so on.

Australian Makers of Compact Tractors

Cox lawn tractors are leading in production of the Australian-made ride-on mowers. Other Australian makers include Greenfield, AMC, Rover, Deutscher and, of course, for push-mowers, Victa. However, the Victa Company does not make any ride-ons or compact tractors.

Cox Industries (Australia) Pty Ltd, Acacia Ridge, Queensland

Owen Cox's company started out in the air-compressor business. Then in 1959 Cox launched the Mowmobile and found a ready market. Today the company he founded makes a wide range of garden tractors. Here is the Cox garden tractor range, made in Queensland, except for the engines:

- The Stockman, in five models, 11.5–14.5 HP B&S front-engine ride-ons with toe/heel forward/reverse friction drive transmissions.

Kingcat Side Discharge Model KR7320 by Cox Industries. 18 HP Briggs & Stratton Vanguard OHV V-Twin engine, Tuff-Torq Hydrostatic Transmission, 40-inch side discharge mower deck. For areas up to five acres.

- The Country, in three models, 14.5–18 HP B&S front engines, friction-drive ride-ons.
- The Z-Cut, a zero-turn rear-engine 18 HP B&S, toe/heel forward/reverse friction-drive transmission.
- The Kingcats, in six models with hydrostatic transmission, 14.5–18 HP B&S engines.
- The Conquest front-mount commercial tractor; 58-inch mower deck with hydrostatic transmission, 25 HP Kohler engine.

Rover Mowers Limited, Eagle Farm, Queensland

Australian-made Rover ride-ons are certified to Australian Safety Standards and come with a two-year warranty.

Rover Mowers Ltd, a Division of GWA International, promotes the advantages of a four-blade cutting disc versus a straight-bar cutter. The difference in price between a set of four swing-back blades and two-bar replacement blades is $16. The bar blades are 70% more

Ranger™	Raider™	Rancher™	Lawn King™

Up to 2ha (5 acres) 7.8kw (10.5 HP)	Up to 2ha (5 acres) 10.1kw (13.5 HP)	Up to 4ha (10 acres) 9.7kw (13 HP) 11.4kw (15.5 HP) 13.4kw (18 HP)	Up to 4ha (10 acres) 12kw (16 HP) 13.4kw (18 HP)

Rover mower range

expensive than the two sets of swing back-blades, and with a straight bar there is a risk of engine damage if the blade hits an obstruction.

Rover Mowers are also the distributors of the Deutscher range of industrial and commercial mowers, including an out-front model, designed and built in Ballarat, Victoria.

Rover four-blade cutting disc with free-swinging blades

Deutscher pedestrian-controlled SP slasher Model H 26, 26-inch cut, with 10.5 HP or 13 HP electric start Honda engine.

Victa Mowers

Victa mowers, as their advertisements famously noted, 'turn grass into lawn'. Victa does not make tractors or ride-ons but the company gets attention with (arguable) claims made that the company founder, Mervyn Victor Richardson, was first in the world to commercialise a rotary mower, so it needs a mention here. Victa Mowers was founded in 1952 and now markets their push-mowers into many countries.

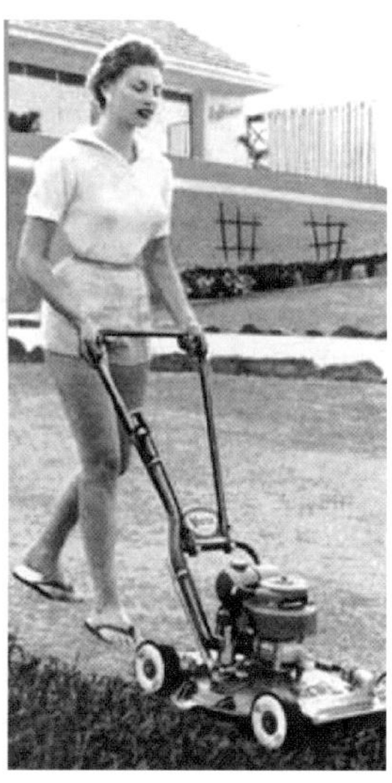

Greenfield's top of the line — the twin Evolution ride-on — is offered with OHV engine options from 16 to 25 HP, 3- inch cutting deck, toe-and-heel foot control on forward motion via a friction drive system. Quick-lock axle differential enables better traction when engaged on slopes or in boggy conditions.

Greenfield Products Pty Ltd

An all-Australian family-owned business established in 1966 and based in Acacia Ridge, Queensland, Greenfield makes ride-on garden tractors. In addition, it makes slashers, domestic mowers, shredders and a range of attachments for ride-ons. An interesting feature of the Greenfield ride-on range is their V-belt variable-speed drive system that enables seamless single-pedal forward and reverse-speed control.

Australian Mower Company

Established in 1993, Horsham-based AMC is another recipient of support under the Victorian government's Grow Your Business programs. AMC's Horsham factory now has robotic welders, automotive-style assembly line and improvements to inventory and warehousing systems to make a range of grass-care equipment.

Mini-skid Loaders: Dingos and Kangas

The stand-on compact utility loader concept originated in Australia around 1980, when plumbing contractor Peter Wiffen built a motorised wheelbarrow. Wiffen subcontracted the Jaden company on Queensland's Gold Coast to build units in response to an emerging demand. Those first units were powered by a 10 HP petrol engine. Demand grew rapidly, especially from equipment rental yards. The Dingo was redesigned as

Above: AMC Explorer XL step-throughs are available in a variety of engine manufacturers and horsepower/deck configurations from 13.5 HP 33-inch deck to 18 HP 40-inch deck.
Top right: The AMC Resort out-front series is available in a similar variety than the Explorer.
Lower right: The AMC Predator rear-engine series have the patented TCS (total control system) which with a simple movement of the steering wheel by the operator controls speed, direction and selection of forward or reverse. They also have foot-operated cruise control for travel once the mower has reached the desired cutting speed. Available in a variety of engine manufacturers and horsepower/deck configurations from 20 HP 45-inch deck to 27 HP 7- inch.

AMC Manor PRO XL. The high seating position gives the operator excellent vision over the front cutter deck and features a fully adjustable suspension seat with comfortable armrests. The flat operator's platform also provides more room and ease of mounting. Features include the hydraulically driven fingertip control of the cutter deck and simple heel-and-toe control pedal for the hydrostatic transmission. Powered by either a three-cylinder 22 HP Yanmar diesel engine or a 25 HP V-Twin Kohler petrol engine, with 60- or 72-inch cutting decks.

a versatile 'hydraulic tool on wheels'. Around 1991 there was a partnership breakup, and Gary Briggs bought the Dingo design to set up a factory in Dalby, Queensland.

At that stage, the remaining partners had an option to buy the equipment back or design another. They did the later, and called it the Jaden Loader. A few years later they changed the name to Kanga Loader when negotiating with American interests. Meanwhile, by 1995 Gary Briggs' Dalby business was burgeoning, including substantial export orders. One shipping container was leaving Australia each week by 1996. He negotiated contracts with Toro in Minnesota to build Dingos for the North and South American markets. The Kanga loader was introduced in the US in 1997.

Gary Briggs remains president of the Australian-

Early Dingo, the Riga'l, 1981–91.

Dingo K9-3 (which replaced the Model 950 in 2002) has a two-pump hydraulic system coupled to the 16.4 kW (22 HP) Kohler engine; 6 km/h top speed; bucket load capacity of 500 kg (SWL 250 kg) and operating weight 745 kg.

based Dingo Mini-Diggers Pty Ltd, registered in Dalby since 1991. Dingo is the market leader in Australia. By 2005 there were more than 4000 Dingos around the country — over 1000 in the hire industry and more than 3000 in the hands of contractors and end-users.

Applications for these mini-skids are ever growing: landscaping, construction sites, internal demolition and high rise construction, trenching, paving, snow removal, fencing, foundries. Attachments are just as diverse: vibrating ploughs, 30-inch tree-planting augers, trenchers, rotor tillers, 4 in 1 buckets, and so on. There are 60 attachments available for these loaders — the equivalent of a Swiss army knife for contractors.

Dingo K9-4 has a two-pump hydraulic system from the 18.6 kW (25 HP) Kohler or 25.6 HP Yanmar Diesel engine, 6 km/h top speed, bucket load capacity of 600 kg (SWL 300 kg) and weighs 900 kg.

The mini-skids are highly manoeuvrable machines, with models as narrow as 800 mm — enough to get through a garden gate — and also fit nicely on a regular car trailer. By 2005 the growing market in Australia was absorbing almost 1000 units a year, one quarter or so into rental yards, with Dingo commanding the biggest proportion of the local market. Power units range from 16 to 25 HP air-cooled engines, both petrol and diesel options.

These mini-loaders spend as much of their time hydraulically powering attachments as they do on front-end loader work. An important outlet for this type of machinery is the equipment rental business, for example, Coates, Kennards and National.

In the US, sales exceed 14 000 units a year and recently the mass-marketers or box stores like Home Depot have entered the equipment rental business, although that has

Dingo back hoe attachment. The machine is less than a metre wide. This takes advantage of the auxiliary 3400 psi hydraulics on the Dingo.

not occurred in Australia.

At the heart of the mini-skid is a complex of hydraulic pumps — up to four on some models, stacked and direct-connected to the engine — that provide power to the ground and all the other attachments or outlets.

Small size and versatility are the key features of mini-skid loaders. They can be operated in walk-behind or stand-on mode. Not having a cab or full panoply of operator comforts keeps costs down as well as facilitating operator access. With increasing sophistication, mini-skids are proliferating in North America, where there are now twelve makers of licensed or similar products. Some models have over 60 attachment options.

Dingo Dumpa.

The Dingo K9-4 is fitted with a remote control package that manages the two drive levers, bucket lift and tilt levers. The remote-controlled Dingo performs like a regular Dingo with the added safety of being operated a safe distance from conditions that are too dangerous for an operator. Remote-controlled Dingos are used in smelters, factories, mines and any place where conditions are not suitable for an operator.

Kanga Model G 724 specifications: engine, 24 HP Honda GX670 (17.7 kW); maximum lift capacity, 250 kg; travel speed, 7 km/h; operating weight including bucket, 857 kg. Shown with Fat Tracks option fitted.

CONCLUSION: THE AUSTRALIAN TRACTOR MARKET ANALYSED

It is possible that when Tractors have outlived their novelty in our midst we may eventually conclude that they have appeared on the scene too late in the development of Australian Farming ever to become an effective part of it.
— *Professor A.J. Perkins, in the* South Australian Journal of Agriculture, *1929*

Professor Perkins' remarks were part of a well-considered economic analysis of the cost of using horses versus tractor farming in 1929. He was quite concerned that if 'tractorisation' continued, Australian family farms would disappear, our land would be all city- or even foreign-owned, and a rural lifestyle and the values associated with it would be gone forever. Tractors were to blame, he asserted.

R. Wanis in National Farmer, *1983.*

Nowadays it would be easy to discount the professor as a gloomy old Luddite because he favoured horses. In 1929 he figured that horses were still economically viable. Perhaps they were, and some of his claims might have proved correct, but there was far more than economics involved.

Do tractors really cause the rural population to decline? Even now, some economists and reporters make such claims; they see a ready scapegoat. People have left the land — rural depopulation and the decline of many country towns is undeniable. But it wasn't tractors that were to blame.

As early as 1911 a McDonald advertisement boasted of the tractor's fuel economy and stated with some foresight: 'it keeps your sons on the farm because it reduces the drudgery and makes farm work more interesting (compared with horse farming)'. Still, that wasn't enough.

Why Do People Leave the Land?

The real cause of people leaving the land is harsh economic reality. The return for labour is significantly lower in farming than in many other professions. Rural life far from towns offers far fewer amenities for the family. If the farmer has children, how is he going to split the farm to support them? Subdividing a broadacre farm can only go so far before it will not be a viable business these days. Another factor is that some young people don't necessarily appreciate country living. Many find rural isolation intolerable, and the noise of the cities a pleasure rather than a distraction.

If young people can leave the farm, many have to do just that. That means reliable labour becomes less available and farm managers are obliged to use labour-saving machinery. Employing staff can create headaches that many independent operators don't want to be bothered with. Machinery and tractor adoption becomes a consequence of the loss or shortage of labour — machinery does not cause people to go.

Harsh Economic Reality

In 1870/71, and again in 1881/82, there occurred two periods of extensive movement away from the wheat-growing areas of Victoria. This was well before advances in labour-saving machinery. For example, McKay's legendary harvester was not marketed until 1886. The self-tying binder with the new Appleby knotting mechanism from the US had not even reached Australian shores. And steam traction engines were of no consequence to Australian agriculture at that time. Harsher living conditions were a factor, but difficult financial circumstances and perceptions of greater rewards for their labour elsewhere made them leave. These still are the primary reasons people leave the land.

In recent decades, with 'tractorisation' complete, there have still been periods of severe rural population decline and drift away from the countryside.

The situation can change or even reverse itself — depending upon prevailing economic circumstances. And the positive impacts of the tractor on Australian farming can be assessed only after the furphy of blaming mechanisation and tractors for rural depopulation have been despatched. Then on the other side of the coin, there are the increasing numbers of 'weekend warriors' — hobby farmers on acreage — and country lifestyle factors to consider.

Not many years back, many farmers found horse farming a drudgery, especially those who were mechanically minded. To them the tractor when it became available was readily accepted. It was the greatest step forward in farming that they had known.

The trend for youth to answer the siren call of the cities has made the use of tractors and machinery essential for those who stayed and were determined to make a go of production farming (as distinct from hobby farming). Conducting field operations at the right time in the season is critical to profitable production. There is a shortage of quality labour in the countryside, thus power farming is the key to timeliness.

Certainly tractors are among the modern farmer's basic tools of trade. Most production farms have from two to four tractors.

The Australian market has always attracted overseas tractor makers. In 1924 Power Farming *included advertising from Case and IHA, as shown here, among others. At least 48 different imported brands have appeared in Australia over the years and there are 39 tractor brands available today.*

Tractor Numbers

According to the Tractor and Machinery Association (TMA), the total number of farm tractors sold over the past two decades has hovered around the 8000 mark. TMA statistics deal with working tractors — tractors that have a drawbar, pto and three-point linkage. Their numbers do not reflect ride-on mowers and outdoor power equipment. That category of tractors and equipment is considered a different market and has been steadily growing. The numbers of compacts, utility and ride-on tractors, front-mounted mowers, skid-steer loaders and all-terrain vehicles has been spectacular and their combined sales exceeded some 50,000 a year in the new millennium; of that number ATV sales have stabilised at around 14,000 a year. These compact machines have considerably enhanced the pleasures of a country lifestyle for increasing numbers of acreage owners, hobby or part-time farmers, and even many urbanites. Compacts, utility tractors and ATV sales are the brightest segment of the machinery market and an important business component of modern dealerships. There are a few Australian brands in the mix. Finally, there are the tractor aficionados who collect or like to see vintage machinery tractors. For them Australia has a large and eclectic mix.

Tractor Nostalgia

In 2001, the greatest number of tractors ever to work in one area simultaneously was 298. The assemblage

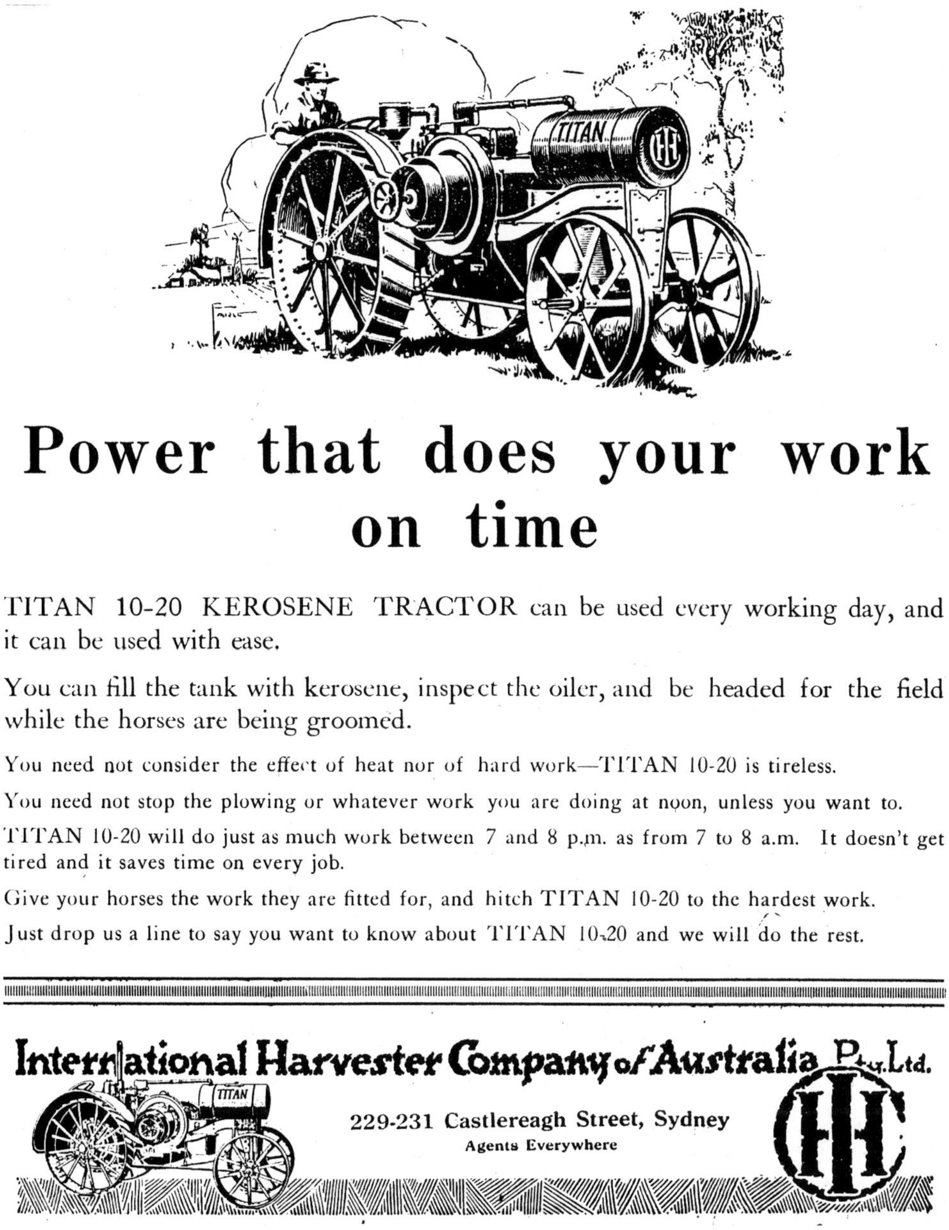

took place in Yass, NSW, at the Easter weekend. That effort made it into the *Guinness Book of Records*. The first tractor to ever come to Australia was there too, the 1903 Ivel.

Not to be outdone, South Africa assembled 730 tractors to work in one field over Easter, 2002. Then Ireland broke that record again by assembling 1832 tractors ploughing in one field at the same time, at Cooley, County Louth, in August 2003. Next England decided to give it a shot. The date was set — 24 August 2003 — and the Great British Challenge was on, at the Hullavington Air Base, Wiltshire. The event was well organised, but sadly they only managed 771 tractors working in the same paddock at the same time. Two of those came from Australia to help.

The latest and greatest assemblage was the Plough and Be Counted 2 exercise that took place on a property called Shaftsbury at Cootamundra, New South Wales, and the *Guinness Book of Records* has registered that gathering as the world record.

The organisers assembled 1914 tractors of all shapes, ages and sizes over Easter weekend, April 2004. However the number had to be scaled back to 1901, since a baker's dozen didn't make the distance around the circuit. So the formal record so far is 1901 tractors. Not only that, but there were a world-record 385 Ferguson TEA tractors in that mixed multitude! International and Fordson Tractor clubs also set out to make records within the record for the most of one brand at Plough and Be Counted 2.

Australian farmers have an advantage when it comes to having the room for such exploits. There are some wheat growers who harvest 60 000 hectares of crop on an individual farm in a season.

Tractors at Cootamundra ranged in age from the 1903 Ivel to the very latest. Two of the tractors that committed to Plough and Be Counted 2 had taken part in all the world record attempts. Their owner also had the machines at Yass, South Africa, Ireland and England, and then came back to Australia for 2004. That in itself could be a record for kilometres travelled by tractors!

Some rare tractors came from England, Germany, Ireland, New Zealand and South Africa to take part in the 2004 world record attempt. Several tractors at Plough and be Counted 2 had to be taken apart, put into containers and reassembled on site. There was an enormous amount of work involved behind the scenes to get this together, so the organisers are hoping the record will stand for quite some time!

Tractor Sales Trends

The graphs that follow are used to draw attention to a number of features of the Australian market for wheeled agricultural tractors:

1. The best year ever for tractor sales was 1964, when 23 422 units were sold.

2. Weather and commodity market fluctuations have been intimately linked with the fortunes of Australian farmers. There were severe droughts in 1943, 1982 and a prolonged dry at the start of the 21st century. There were the wool and wheat booms in 1953 and 1964, respectively. Other factors that have affected tractor market volatility have included government incentive schemes, tariffs and tax concessions.

3. In more recent years, following drought, recession and a severe decline in farm income between 1983 and 1985, annual tractor sales slumped, so that by 1987 they were approaching the dismal levels of the 1930s. Even now they are not much better in terms of unit sales — but nowadays the total power sold and dollar value is much higher — and holding.

4. According to *Commonwealth Yearbook* statistics (which have unfortunately been discontinued in this area) the total number of tractors on farms peaked in 1978 and since around 1990 has gradually declined.

5. At least 48 overseas makers are known to have placed tractors on the Australian market over the past few decades. Currently there are around 39 brands from no less than 13 countries, each vying for their piece of the Australian tractor market.

6. Tractor dealers and agencies have closed down at such a rate that an estimated 30 000 jobs in tractor sales and service were lost from the business between 1960 and 1987 (Dunn, 1987, personal communication).

7. Four-wheel-drive tractor sales (equal sized wheels) peaked at about 1200 units per year in 1984 but have trailed off to very low levels in the past few years. All makers, both local and foreign, have moved away from this market and some have gone out of business, or merged with corporate rationalisation.

8. A recent phenomenon is the emergence of that class of tractors with unequal sized drive wheels, or front-wheel assist (FWA). The combined number of tractors sold with FWA and FWD now exceeds

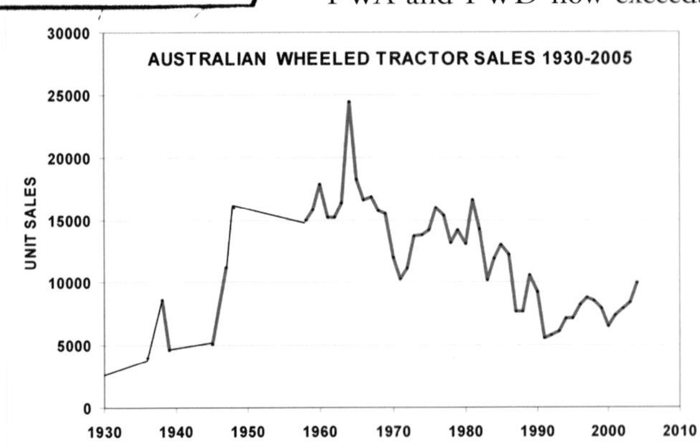

Tractor unit sales in Australia, 1936–2004. (Data derived principally from Power Farming *articles, BAE and TMA statistical reports). Some articles have reported 1951 as the peak sales year for agricultural tractors, but the author has not been able to verify that more tractors were sold then than in 1964, when annual sales peaked at 23 422 units. The worst years in recent memory have been 1987, when sales slumped to 7 276 units, then 1991/92 at around 5 700 tractors (TMA figures).*

two-wheel-drive in all size ranges, according to the Tractor and Machinery Association.

9. Sales of tractors of less than 15 kW (20 HP) have increased strongly, whereas all other segments of the market have been variable, depending on demand. These compacts are mainly used for industrial, recreational, part-time farming or landscaping pursuits. The distinction between rural users and others, and between outdoor power equipment in the smaller categories is blurred. However, some compacts are used in broadacre work for various farm tasks like slashing and fencing.

10. The Australian farm tractor market is largely a replacement market nowadays. On the other hand the compact/outdoor equipment business is still growing.

Australian Tractor Makers' Market Share

Tractor sales increased remarkably after World War II, up until 1964. Australian manufacturers enjoyed a share of the market for a time, but no longer.

While local manufacturers never enjoyed the major share of the Australian farm tractor market, their share was probably highest in the 1960s, at around one third of the total market.

Today it has simply vanished.

The total Australian market of around 10 000 units in 2004 is small, up from previous years but it is now shared by some 39 foreign makers, each with a large number of models — as many as 80 in some brands — so there are more than 600 models of new tractors on offer. Unit sales of even a popular individual model might not exceed 50 a year. The numbers are no longer sufficient to justify gearing up for local production, using conventional production technology and pricing structures.

The Australian market has always attracted overseas makers, as it provides a buffer for them against seasonal and cyclic sales trends in the Northern Hemisphere, where their seasons are the reverse of ours. There have been almost 50 different makes competing for this market over time, even though Australia has represented less than one percent of the world's tractor market.

Local makers once received tariff support protection, but that bounty was removed in 1986. One consequence was that Chamberlain stopped production in Western Australia. Export incentives have, however, been valuable for a handful of local makers of self-propelled equipment, but their relative volumes are tiny. The brutal facts are that local tractor manufacture has not been sustainable because the market here is too small, we are too far away from the export markets, development costs are too high here, and the sheer cost of maintaining parts and service with so many models today is beyond the locals.

Tractor Power Trends

Until 1985, there seemed to be a certain inevitability (which had been evident for four decades) about the increase in average power of tractors sold in Australia. Since 1985, however, the trend has been slowed. Three factors in particular account for this:

1. The number of four-wheel-drive (equal sized wheels) tractors declined out of proportion to total sales. They were replaced by front-wheel-assist (unequal sized wheels) tractors of somewhat lower power, although that has been edging up steadily in the past decade.

2. The oftentimes stagnation of world agricultural commodity prices — pressed especially hard between 1986 and 1988, with subsequent declines in output of the cropping industries — forced farmers to reassess their approaches to production. This led to the increased adoption of such practices as minimum or zero tillage and more recently precision farming — and add controlled traffic farming — in order to reduce input costs.

3. Some of the cycles in agriculture had graziers, and specialist crop farmers, providing the mainstay of demand for new tractors. Their needs are usually for smaller tackle. Crop farmers who had the big machinery simply did not trade tractors during commodity downturns.

The rapid adoption of front-wheel-assist (first on the smaller sized tractors) has been influential in the shift to lower power tractors. FWA enables more efficient use of engine power under less favourable ground conditions, enhances tractor safety, and is available at reasonable cost compared with FWD tractors with equal sized wheels. Incidentally, front-wheel-assist tractor axles have never been made in Australia. Finally there has been the shift to minimum tillage, controlled traffic and other cost-minimising farming techniques. These have reduced the need for the higher power or the total number of tractors. As for our 4WD component-build tractor makers, they quietly passed on when Waltanna's last one left the factory in 1991. At the other end of the power spectrum is the inflow of Asian-built tractors, with China, Korea and India the latest and most aggressive on the Australian market. More astounding has been the rapid rise of outdoor power equipment sales — mowers, compact ride-on tractors, front mowers, mini-skidsteers and ATVs.

The Australian Wheat Machine

An instructive exercise is to look at the efficiency of Australian agriculture at producing our foremost farm commodity, wheat. How does wheat production output compare with aggregate annual tractor power input, using tractor sales as a rough guide?

The first graph on the next page was generated by dividing total wheat production by annual tractor aggregate power based on new tractor sales, to show the

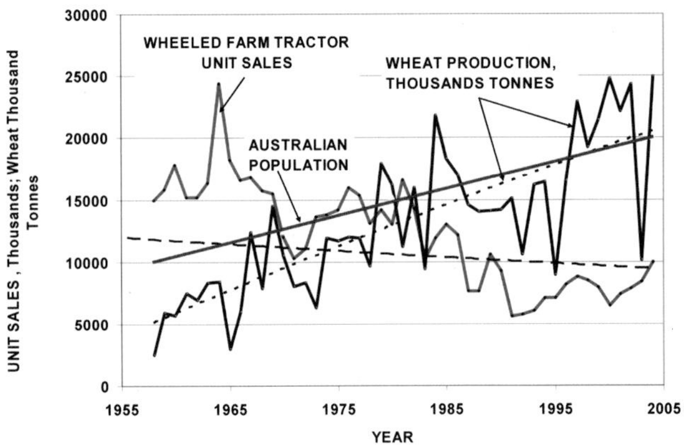

A closer look at total Australian wheat production compared to tractor sales between 1958 and 2004. Wheat has been the backbone of the cropping economy.

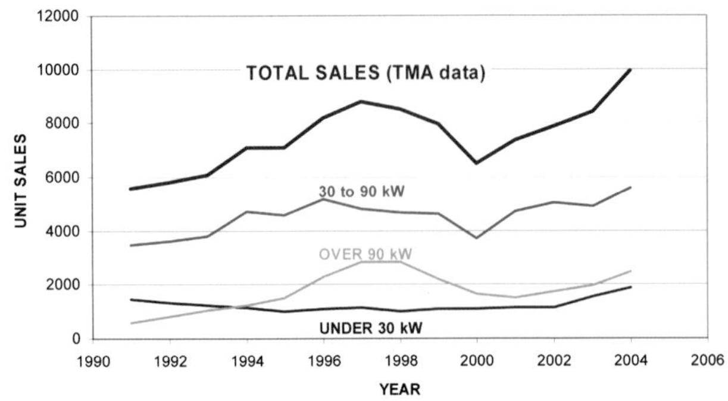

Wheeled tractor sales, 1991–2004, broken down according to tractor power classes.

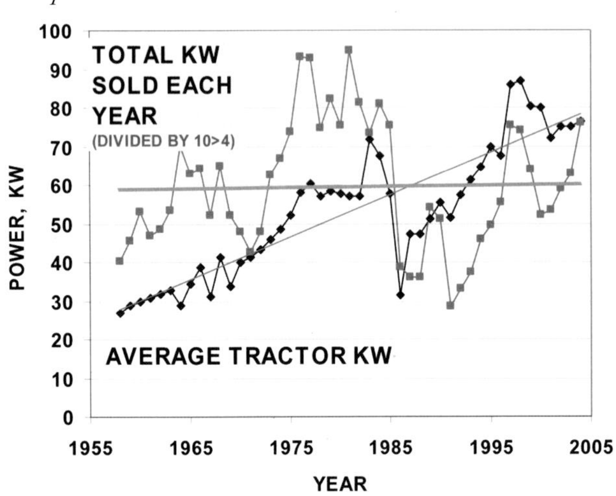

Aggregate and average wheeled agricultural tractor power sold in Australia each year, 1958–2004 (from TMA data).

number of tonnes of wheat produced per kilowatt of tractor power sold for a given year. This is obviously not a precise way to assess efficiency but the general trend lines from the data are useful.

Between 1961 and 1981 the Australian 'wheat machine' produced about 15–17 tonnes of wheat for each kilowatt of tractor power sold. Then came the oil crisis around 1979–80. Strangely enough, the tonnes/kW fell at first, perhaps indicating that fuel price, which rose sharply at that time, was not a major determinant in making the wheat production machine any leaner. Then came a dramatic change after the drought and rural recession of 1981. By 1986 wheat growers were generating more than twice as much grain for their investment in tractor power (over 35 tonnes wheat per purchased kilowatt). In these lean economic conditions, cutting input costs is the key to economic survival. Wheat growers have done that by changing farm practices. As for crop commodity prices, they have not changed substantially and are controlled offshore by international trade in any case. The recent drift has been to standardising around 12 metre wide for tillage and seeding equipment — or modules of 12 metres to match grain head widths.

Value for the Dollar Invested

The investment allowance went in the 1980s, and depreciation allowances are tighter, so there are incentives to buy secondhand tractors.

The average retail price of wheeled tractors in Australia escalated in the 1960s — especially with the growing emphasis on operator safety and comfort. No longer was it acceptable to throw a wheat bag over the pan seat and go out to work exposed all day to the elements. With increased sophistication came higher prices, as tractors were more uniformly equipped with cabins with air-conditioning (refrigerant-type aircon), improved hydraulic systems, including remote couplings, ground speed and live pto standard, gear shift ratios changeable on the move, and now continuously variable transmissions (CVTs) and power steering, etc. Today's broadacre tractors are also packed with electronics, but they come at a price.

Net tractor weight per HP has not altered significantly, so the hike in price of tractors is not due to more steel but to an aggregation of better quality, higher costs of marketing and backup, and increased sophistication, especially in electronics. Tractors have also become considerably more fuel-efficient, but that too has come at a price. All tractors use diesel power units. Many of the smaller tractors and all the larger powered tractors are turbocharged, with many aftercooled for greater fuel efficiency.

Summary

Some 25 local manufacturers have attempted to make a business out of building tractors in Australia. None have survived.

A dozen makers of self-propelled machines have had a very small piece of the local market and a handful survive with some innovative machines.

There have been no Australian breakthroughs in tractor design, although there were many initiatives

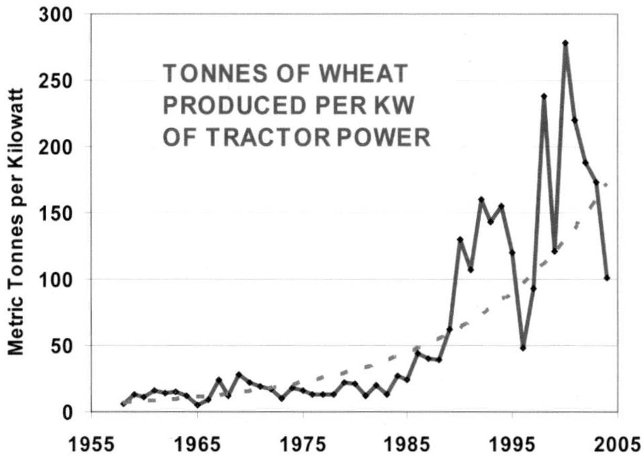

Tractor cost over the years, in terms of the average kilowatts sold compared with the average price of tractors over the decades. Note that tractor power and weight are very closely related. While power to weight ratios have stabilised in recent years, tractor costs have escalated compared to commodity prices.

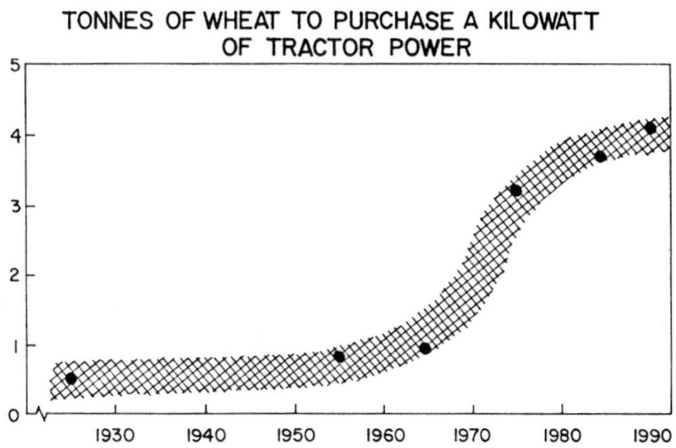

The ratio of wheat production year by year divided by purchased tractor power through each of those years. The Australian wheat machine has improved in efficiency. The methodology may be questionable and can be skewed by certain factors, but the trend is highly visible.

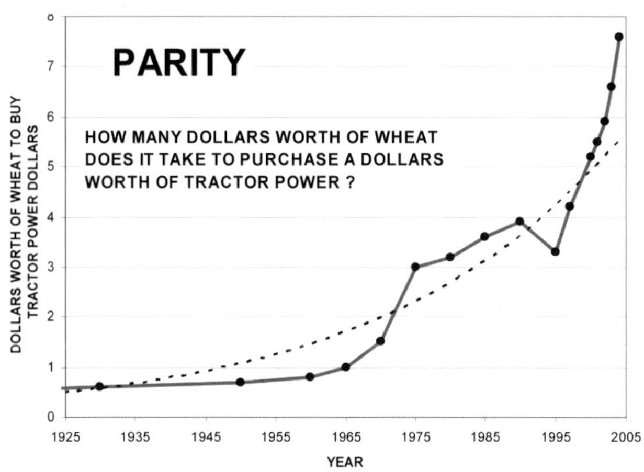

Parity index. The amount of tractor power to produce a tonne of wheat has declined, but on the other hand the number of dollars worth of wheat required to buy tractor power has escalated with time. Simply put, commodity prices have been slack while tractor costs have been increasing relentlessly.

aimed at this turbulent, isolated and, by world standards, comparatively small market.

There have been no tractor breakthroughs, but some important advances in electronics and controls, such as the Beeline Auto-Steer pioneering work.

For offshore farm machinery makers, Australia has been and still is a valuable testing ground: it gives them a place to test in a country where there is a common language, willing farmers and another season in their year (opposite to the Northern Hemisphere).

The market share for a given tractor model is simply too small for local tractor production.

Dealerships have shrunk and many surviving dealerships have become multi-branded. New tractors are scarce on the showroom floors of the few service-oriented dealerships. Prospective purchasers of a new tractor in today's market have to plan ahead for delivery.

The compact power equipment business is now an important source of additional work and a highly important source of income for tractor dealers. And in that area several Australian makers are active.

In 1988 W.T. Brown wrote: 'There are no technological or economic arguments to support local manufacture, but the loss of engineering expertise with the closure of factories and the reduction of staff by importers will affect the quality of tractor servicing, and the preparation and scrutiny of regulations and standards to the detriment of the agricultural industry.'

However, pioneer tractor designer Bob Chamberlain should be allowed the final word on Australian tractors: 'Manufacturing costs in Australia have greatly increased ... We seem to have developed, then lost, our ability to design and build ... tractors in Australia. Some years ago I was invited by Sochiro Honda to Japan as his guest. Throughout his organisation and at all levels, the workers would be talking about "Honda methods" — hard work and loyalty to their employer. Australia is well placed to make bricks and beer competitively — but maybe this is all many people here want to make' (A.H. Chamberlain, personal communication, 1986).

BIBLIOGRAPHY

Books and Articles

Agricola, *The Man on the Land*, W.K. Thomas and Co., 1910.

American Society of Agricultural Engineers, *Agricultural Engineers Yearbook*, ASAE, 2004.

American Society of Mechanical Engineers, *The Hart-Parr Tractor,* ASME Historical Heritage Landmark, 1996.

Atkinson, J., *An Account of the State of Agriculture and Grazing in New South Wales*, London, 1826.

Baillie, W.F., and G.H. Vasey, 'Rating and Testing of Stock Model Goods', *Journal of Australian Industrial Engineers*, Vol. 40, 1968.

Baillie, W.F., 'The Changing Trends of Tractors in Agriculture — Past, Present and Future', *Power Farming Technical Annual,* 1969/70.

Baillie, W.F., and G.H. Vasey, 'Graphical Representation of Tractor Performance', *Journal of Australian Industrial Engineers*, Vol. 41, 1969.

Baillie, W.F., and W.T. Brown, 'Summary of Tests of Tractors in the 60-70 HP Class', *Power Farming*, 1972.

Baldwin, N., *Kaleidoscope of Farm Tractors*, Frederick Warne, 1978.

Birmingham, J., I. Jack and D. Jeans, *Industrial Archeology in Australia: Rural Industry*, Heinemann, 1983.

Birmingham, J., I. Jack and D. Jeans, *Australian Pioneer Technology,* Heinemann, 1979.

Blainey, G., *The Tyranny of Distance*, Macmillan, 1968.

Blyth, K.A., 'A Wool Grower in the Cane Fields', *Australian Canegrower*, December 1983 – February 1984.

Bonnett, H., *Saga of the Steam Plough*, George Allen and Unwin, London, 1965.

Bonnett, H., *Traction Engines*, Shire Album 143, Shire Publications, 1985.

Bromby, R., *The Farming of Australia*, Doubleday Australia, 1986.

Brown, W.T., 1988, 'The Australian Tractor Industry in 1988', *Transactions Mechanical Engineering Division*, Institute of Engineers, May 1988.

Burke, B., and J. Norris, *The Australian Steam Engine Listing*, Sydney, 1978.

Burrows, G., and C. Morton, *The Canecutters*, Melbourne University Press, 1986.

Caterpillar Tractor Co., *Fifty Years on Tracks*, Caterpillar, 1954.

Cawood, C., 'The Early Days of Farm Tractors', *The Automobile*, June 1964.

Cawood, C., *Vintage Tractors*, Shire Publications, 1980.

Clymer, F., *Album of Historical Steam Traction Engines and Threshing Equipment*, Stemgas Publishing, 1959.

Conigrave, J.F., *A Description of the South Australian Seedsower, Stripper and Winnower*, Adelaide, 1883.

Darwin, N., *The History of Ford in Australia*, Eddie Ford Publications, 1986.

Deere & Co., *John Deere Tractors, 1918–1976*, John Deere Distribution Service Centre, 1976.

Dunsdorfs, E., *The Australian Wheat-Growing Industry, 1788–1948*, Melbourne University Press, 1956.

Edgington, D.W., and C. Hudson, *Stationary Engines for the Enthusiast*, Edgington & Hudson, 1981.

Engineer, 'The Evolution of the Chain-Track Tractor', *The Engineer*, August and September 1917.

Ford, H., 'Henry Ford and the Farm Tractor', *The Farm Journal*, October 1915.

Fream, W., *Elements of Agriculture*, 8th edition, London, 1911.

Fussell, G.E., *The Farmer's Tools*, Orbis Publishing, 1981.

Gray, R.B., *The Agricultural Tractor, 1855–1950*, revised edition, American Society of Agricultural Engineers, 1975.

Griffiths, G.N., *Some Northern Homes of New South Wales*, The Shepherd Press, 1954.

Haining, J., and C.R. Tyler, *Ploughing by Steam*, Model and Allied Publications, 1970.

Hayward, A., 'Steam Power in Agriculture in New South Wales', unpublished thesis, University of Sydney, Department of Archeology, 1985.

Hermann, K., *Ackergiganten: Technik, Geschichte und Geschichten*, Westermann, 1987.

Hermann, K., *Traktoren in Deutschland, 1907 bis heute: Firmen und fabrikate*, DLG-Verlag,1987.

Holbrook, S.H., *Machines of Plenty: Chronicle of an Innovator in Construction and Agricultural Equipment: J.I. Case*, revised edition, Macmillan, New York, 1976.

Hordern, A., *Catalogue of Exhibits*, Easter Show, Sydney, 1914.

Howard Cultivators Ltd., '"Howard" Cane Harvester — Original Report of Test', *Power Farming in Australia,* November 1929.

Hughes, W.J., *A Century of Traction Engines*, Pan Books, 1972.

Hutchings, M.J., and G.H. Vasey, 'Preliminary Investigations of the Noise Levels Around Agricultural Tractors', *Journal of Agricultural Engineering Research*, 1964.

Irvin, E., *Early Inland Agriculture: Farming in the Southern Districts of New South Wales*, Wagga Wagga, 1962.

International, *Farm Power*, 2nd edition, International Harvester Co. of America, 1928 (reprinted by American Society of Agricultural Engineers).

International, *Muscles or Motors?*, International Harvester Co. of America, 1962 (reprinted by American Society of Agricultural Engineers).

Janssen, J.E., *Hart-Parr Tractor's Contribution to the Advancement of Agriculture*, ASME, 1996, www.asme.org/history/brochures/h190.pdf.

Johnson, P.C., *Farm Inventions in the Making of America*, Wallace-Homestead Book Co., 1976.

Keith, T.B., *The Horse Interlude*, University Press of Idaho, 1976.

Kendall, F.J., *H.V. McKay — Pioneer Industrialist*, Monograph 2, Science Museum of Victoria, 1979.

Kenyon, A.S., 'Clearing by Traction Engine', *Journal of Agriculture (Victoria)*, May 1908.

Kenyon, A.S., *Report of the Engineer for Agriculture ... for the Years 1907–10*, Victorian Department of Agriculture, 1910.

Kjellstrom, B., *Biomass Gasifiers for Energy Supply to Agriculture and Small Industry*. Ambio, 1985.

Land, Transport and Sea Trade, 'Types of Tractors Used on Australian Farms', *Land, Transport and Sea Trade*, 1919.

Lairsen, L., *Farm Tractors, 1950–1975*, American Society of Agricultural Engineers, 1981.

Lindsay, J., 'The Farm Equipment Industry — a Necessary Restructure', *Journal of the Agricultural Engineers Society (Australia)*, 1987.

Linge, G.J.R., *Industrial Awakening: A Geography of Australian Manufacturing, 1788–1890,* Australian National University Press, 1979.

Loudon, J., *Encyclopaedia of Agriculture*, Longman, Orme, Brown, Green, & Longmans, 1839.

Loyau, G.E., *The Gawler Handbook*, Goodfellow and Hele, 1880.

Manby, T.C.D., *Interpretation of Tractor Test Reports by the User*, FAO Information Working Bulletin 19, 1960.

Maslin, R., *Big Lizzie: The Story of a Man and a Machine,* The Sunnyland Press, 1982.

McCann, J., *Victorian Steampower: An Illustrated Social History*, Victoria Press, 1994.

McDonald, K.N., *A.H. McDonald, Industrial Pioneer*, privately published, 1988.

McKinley, M., *Wheels of Farm Progress*, American Society of Agricultural Engineers, 1980.

Miller, A.S., I.M. Mintzer and S.H. Hoagland, *Growing Power Bionergy for Development and Industry*, Study 5, World Resources Institute, April 1986.

Milner, P., *An Assessment H.V. McKay's Harvester Works at Sunshine Relative to Other Engineering Manufacturers in Victoria*, Melbourne, 1987.

Morton, C., 'The First 90 Years of Mechanical Cane Harvesting in Australia', *Australian Canegrower*, August–December 1979.

Norbeck, J.C., *Encyclopedia of American Steam Traction Engines*, Crestline, 1976.

New South Wales Department of Agriculture, *The Farmer's Handbook*, 1911.

New South Wales Producer Gas Research Committee, 'Is Charcoal Gas the Solution of the Wheat Farmer's Problem?', *Agricultural Gazette of New South Wales*, 1940.

Perkins, A.J., 'A Critical Glance at Some Aspects of Tractor Farming in Australia', *South Australian Journal of Agriculture*, 1929–30.

Port, L., and B. Murray, *Australian Inventors*, Cassell Australia, 1978.

Posthumus, C., *The Story of Veteran and Vintage Cars,* Hamlyn, 1977.

Queensbury Group, *The Book of Key Facts*, Cassell Australia, 1978.

Quick, G.R., and W.F. Buchele, *The Grain Harvesters*, American Society of Agricultural Engineers, 1978.

Quick, G.R., 'The Australian Tractor Industry: Historical Perspectives', Conference on Agricultural Engineering, Hawkesbury, NSW, 1988.

Quick, G.R., *Two Hundred Years of Farm Machinery in Australia*, New South Wales Department of Agriculture, 1988.

Rennie, E.J.C, 'Motor Tractor Trials', *Journal of Agriculture (Victoria)*, 1918.

Royal Agricultural Society of Victoria, 'Motor Tractor Trials', *Journal of Agriculture (Victoria)*, December 1918.

Schubert, J., *Cab, Sir? The Story of Frank Gason*, A.F. Gason, 1987.

Shearer, D., *Seventy-Five Years of Progress*, Shearer & Co., 1952.

Simpson, M., and P. Simpson, *Old Farm Machinery in Australia: A Field Guide and Source Book*, Kangaroo Press, 1988.

Society of Automotive Engineers, *An Historical Perspective on Farm Machinery*, SAE Publications, 1980.

Spence, C.C., *God Speed the Plow*, University of Illinois Press, 1960.

Stephens, H., *The Book of the Farm*, W. Blackwood, 1891.

Stow, J.P., *South Australia, its History, Production and Natural Resources*, Government Printer, Adelaide, 1883.

Vasey, G.H., and W.F. Baillie, 'Tractor Testing in Australia', *Journal of Agriculture (Victoria)*, June and July 1960.

Vickers, 'The Vickers-Aussie Tractor', *Implement and Machinery Review*, December 1925.

Wadley, D.A., 'Corporate Decision-Making During Recession', unpublished PhD thesis, Australian National University, 1974.

Wadley, D.A., 'Tractor Technology: Progress and Prospects', *Journal of the Australian Institute of Agricultural Science*, 1983.

Watson, C. (Ed.), *Boots and All: An Oral History of Farming in Victoria*, Friends of the Earth, 1984.

Wear, C., *Australian Technology in Advertisements*, NSW Steam Preservation Co-Operative Society Ltd, 1981.

Weickhardt, G.G.T., 'Victoria's Engineering Heritage First Hundred Years, 1842–1942', paper presented to the Second National Conference on Engineering Heritage, May 1985.

Wendell, C.H., *150 Years of International Harvester*, Crestline Publishing Co., 1981.

Wendell, C.H., *Encyclopedia of American Farm Tractors*, Crestline Publishing Co., 1979.

Wendell, C.H., *Nebraska Tractor Tests Since 1920*, Crestline Publishing Co., 1985.

Wheelhouse, F., *Digging Stick to Rotary Hoe: Men and Machines in Rural Australia*, Cassell Australia, 1966.

White Farm Equipment, *Progress in Tractor Power From 1898*, White Farm Equipment Co., 1975.

Wik, R.M., *Steam Power on the American Farm*, University of Pennsylvania Press, 1953.

Wik, R.M., *Tracks and Combines*, American Society of Agricultural Engineers, 1984.

Williams, M., *Steam Power in Agriculture*, Blandford, 1977.

Williams, M., *Great Tractors*, Blandford, 1981.

Williams, M., 'Nostalgic Days of Steam', *Power Farming*, 1978.

Wilson, Bert, 'Bert Wilson Sets up Shop', *Farm Mechanics*, February 1939.

Serial Publications and Useful Websites

- www.tomm.com.au *The Old Machinery Magazine* (TOMM) website.
- http://www.ytmag.com/articles/artint208.htm
- http://members.ozemail.com.au/~bobkav/

aussietractorpage/aussietractors.htm Bob Kavanagh's condensed story on Aussie tractors
- http://www.toytractorshow.com/blue_steigers_ii.htm
- http://shop.screensound.gov.au/prod3479.htmimag
- www.RallyBadges.com
- www.ploughbooksales.com.au Comprehensive collection of books on offer from Plough Book's catalogue
- www.blueflyer.com.au
- www.machineryshed.com
- www.fhec.co.uk *Farm and Horticultural Equipment Collector Magazine*, Kelsey Publishing Group
- www.Stationary-Engine-Magazine.co.uk Kelsey Publishing Group
- www.cpmmag.co.uk *Classic Plant and Machinery Magazine*, Kelsey Publishing Group
- www.oldtractor.co.uk *Old Tractors Magazine*, Kelsey Publishing Group
- www.agsites.net/ Claims to be the world's largest agriculture and outdoor equipment directory
- www.ag-museum.mb.ca Manitoba's Agricultural Museum
- *The Agricultural Gazette of New South Wales*
- *Agricultural History*
- *The Australasian Agriculturalist*
- *Australasian Hardware and Machinery*
- *The Australasian Ironmonger*
- Australian Bureau of Statistics: *Sales and Stocks of New Tractors* (Ref. 1301.0. Canberra)
- *Australian Town and Country Journal*
- *The Farm Journal*
- *Farmer and Grazier*
- *Official Yearbook of the Commonwealth of Australia*
- The Olde Machinery Mart, Townsville
- *The Old Machinery Magazine*
- *Power Farming* (first published 1924)
- *Power Farming Technical Annual* (commenced 1937)
- PROFI International www.profi
- *Statistical Register of New South Wales*
- Tractor and Machinery Association of Australia, *Quarterly Reports* www.tma.com.au

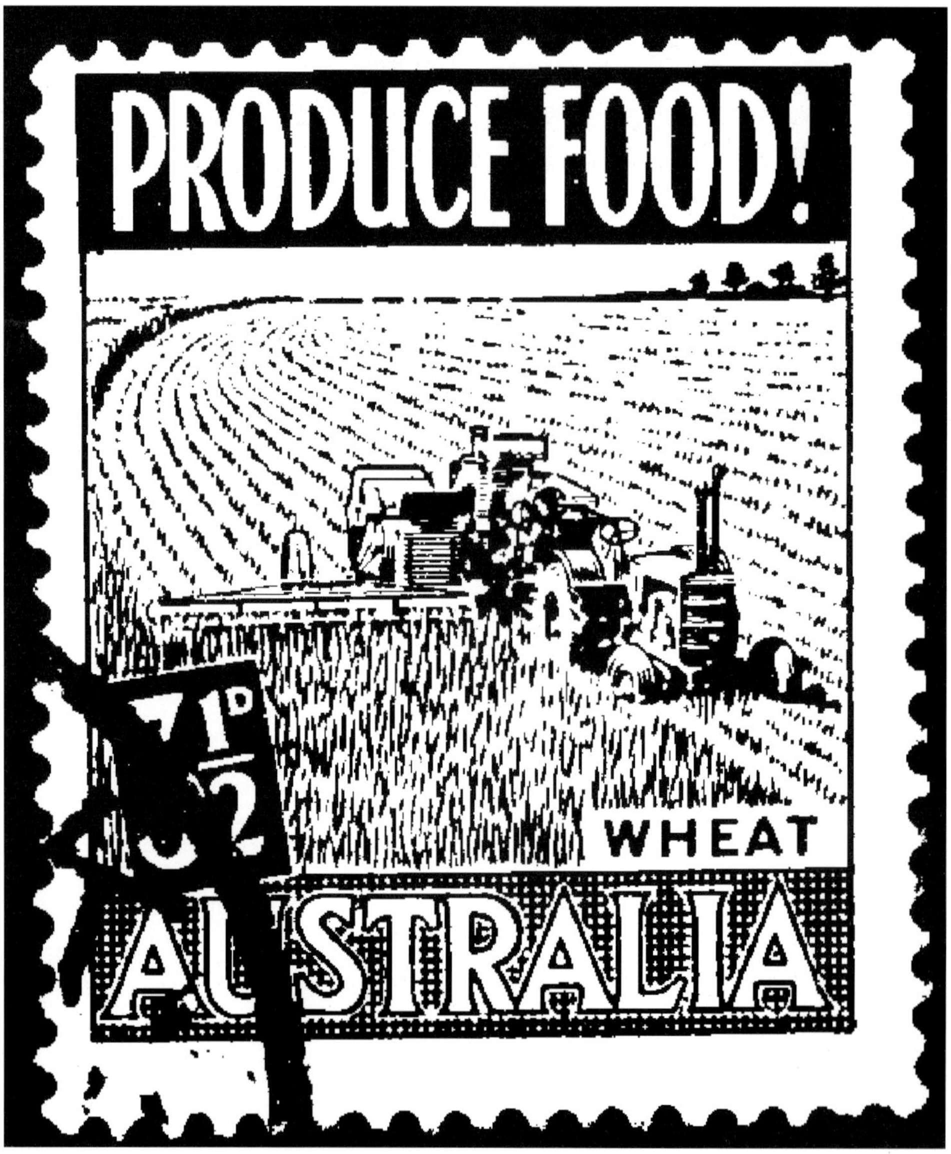

One of two known Australian stamps featuring a tractor. Some countries which have never had a tractor industry, for example New Zealand and Mongolia, have issued numbers of stamps showing tractors and farm machinery. Australia's highly-urbanised population would go hungry without farm tractors.

GALLERY OF LINE DRAWINGS

When I was working in the Philippines (1988-1995) as Head of the Agricultural Engineering Division at the International Rice Research Institute, I commissioned and paid several Filipino graphic artists to do a number of line drawings of machines. The drawings were based on photos that I gave them to work from; and I did some drawing myself on others derived from trade literature. Here is a sampling.

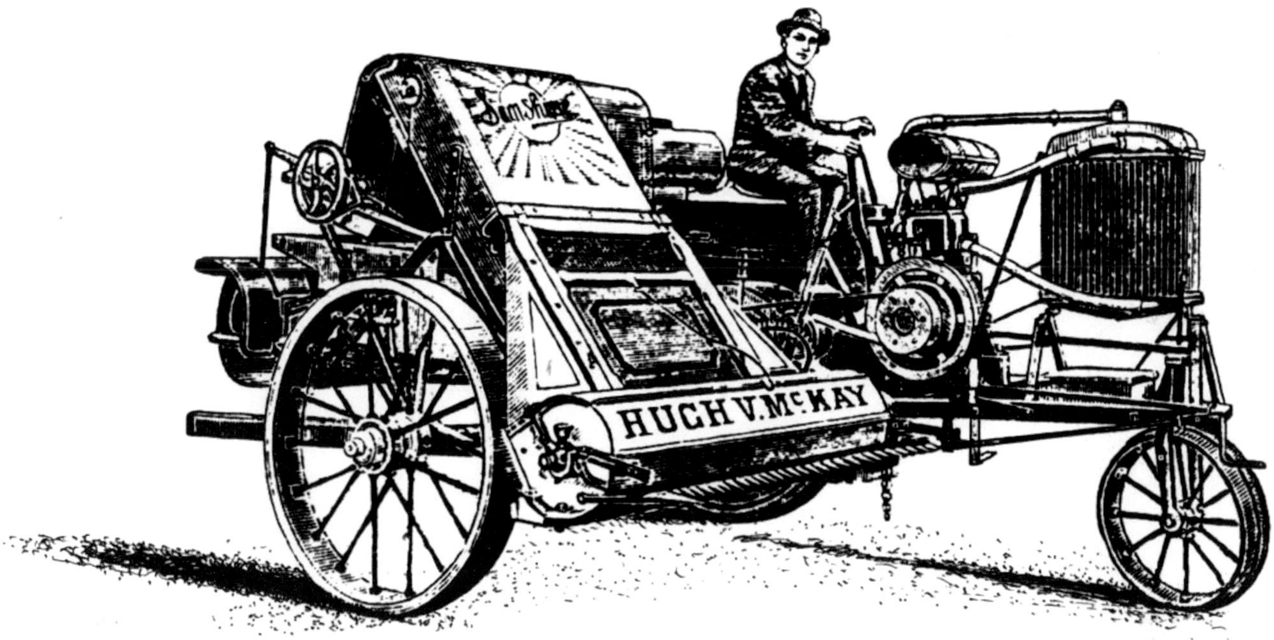

'Sunshine Horseless Harvester'. H.V. McKay experimented with Self-Propelled Stripper-Harvesters between 1907 (first patent) and 1920. He tested several of them off-shore, one even as far away as Spain, but never took them to serial production. Note the single-lever steering. The plan was that the engine could be removed to use on other farm jobs between harvests. They were superseded by the *Sun Auto Header* a few years later.

Frank Bottrill's *Dreadnaught wheels* or *'Pedrails'* were fitted to this tractor and trailer to improve floatation on sandy or difficult country. This outfit was manufactured by Austral-Otis specifically for the Mt Gunson copper mines in 1911. It was powered by a McDonald 45 HP engine.

Victorian farmer Thomas Quinlivan patented a four-wheel drive steam traction engine in 1908 and Cowley's Eureka Iron Works started building several but these were completed under contract in 1910 by Johnson's Tyne Foundry, South Melbourne. The steam engine was mounted under the boiler and the closed steel wheels were used as water tanks. The machine was used for transportation and land clearing and was power steered.

McDonald's 20 HP *Steel Horse* , manufactured in 1916 to try and match American tracked tractor imports and exhibited at the 1916 Melbourne Royal Show as the *"British Juggernaut's Baby Brother"*. Redgum wooden blocks were bolted to the steel track for traction and floatation. It was never produced in quantity.

McDonald Imperial Oil Tractor, the 'Scrub Pusher' version of the Model EB, in 1913.

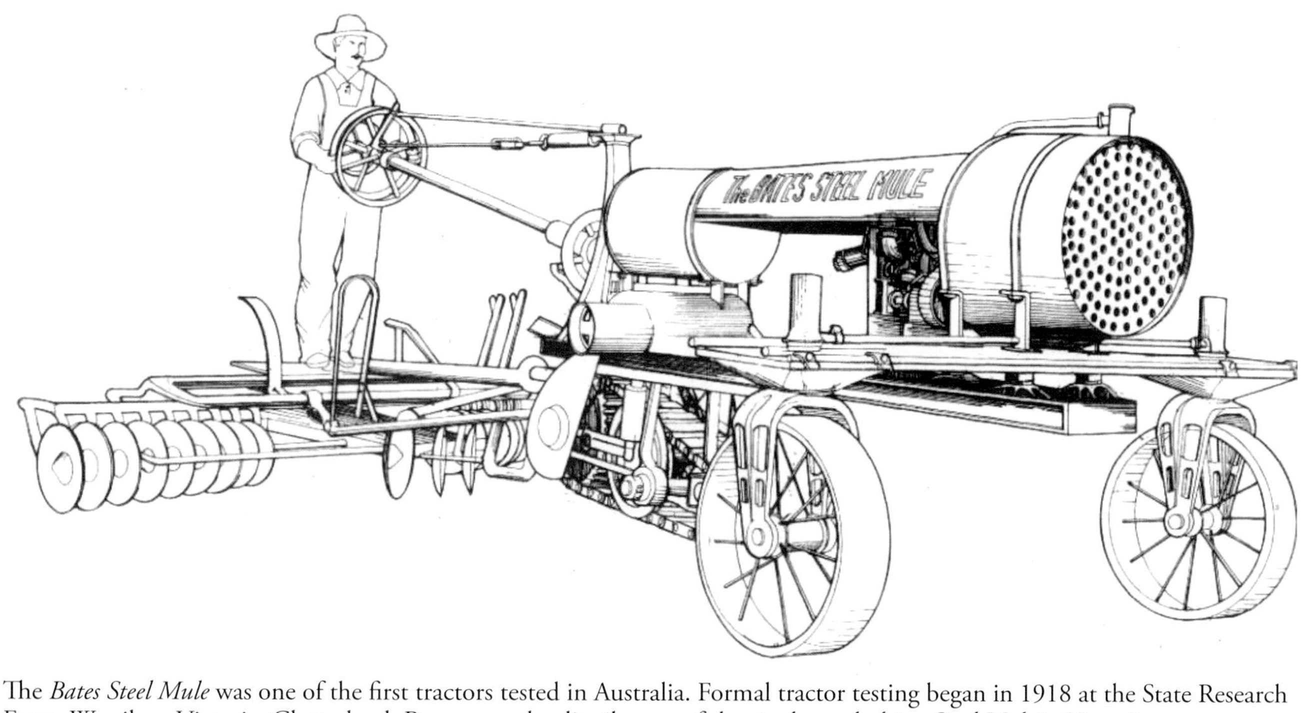

The *Bates Steel Mule* was one of the first tractors tested in Australia. Formal tractor testing began in 1918 at the State Research Farm, Werribee, Victoria. Clutterbuck Bros. were the distributors of the single-track drive *Steel Mule* in Victoria.

Chamberlain 40K, 1945-1955. The kerosene-fuelled 40K had a locally-designed twin cylinder horizontally-opposed engine of 41 HP at 1200 rpm. The 40K had nine forward speeds gears, which was more gear ratios than the competition at the time.

McCormick International AW-6. The 38 HP kerosene-powered AW-6 was International Harvester Australia's first locally-built tractor. IHA had plans to build tractors at their Geelong, Victoria, factory in 1938. The entire machine, engine and all, was locally-made and it finally rolled off the assembly line in 1949. By 1955 they had made 10,000 tractors and IHA was offering eleven different tractor models.

LIST OF COLOUR PLATES

Plate 1 (facing page 72)
Top: The first commercial Australian-built tractor, an Imperial Oil Tractor Model EA by A.H. McDonald & Co., was exhibited at the Royal Melbourne Show in 1909. It operated on kerosene. (Photo courtesy *The Old Machinery Magazine*, February/March 2005)

Bottom: Cowley Steam Traction engine made by Cowley's Eureka Iron Works, Ballarat, around 1913. This was one of the last steamers made in Australia. (Illustration courtesy Martin Hallett, Museum of Victoria)

Plate 2 (between pages 72 & 73)
Top: Norm McKenzie of Cumnock, NSW, owns this 1903 Ivel tractor. There are several Ivel tractors in Australia. Norm McKenzies' is serial number 269. It is one the oldest operable tractors in the world. (Photo courtesy Val McKenzie)

Bottom: In April 2003, a total of 1914 tractors was mustered on a farm near Cootamundra, NSW, establishing a world record for the greatest number of tractors operating at the same time in one place. The *Guinness Book of Records* has acknowledged this as the largest gathering of tractors at work. Actually 13 tractors did not make it around the circuit so the recorded number is 1901. McKenzies' Ivel was in this group. Some 23 charities benefited from the 2004 assemblage. (Photo courtesy Vic Muscat)

Plate 3 (between pages 72 & 73)
Reg Schuster of Palm Beach, Queensland, has restored a Caldwell-Vale motor truck. Here it is after restoration. Reg shows it regularly at the South Coast Restoration Society's Annual Rally. Inset shows the truck as found. The Caldwell-Vale company built Australia's first four-wheel-drive tractor similar to this truck. (Photos supplied by Reg Schuster)

Plate 4 (facing page 73)
Top: 1920 Jelbart owned by Jack Burrell of Gilgandra, NSW.
Bottom: The Ronaldson-Tippet 20-36 SuperDrive Tractor on rubber in 1937. (Courtesy *The Old Machinery Magazine*, Aug/Sept, 2004)

Plate 5 (facing page 88)
Top: McKay Stripper Harvester, 1884 and Hugh Victor McKay, as seen on a centenary celebration first day cover issued by Australia Post in 1984. MacKay went on to develop a multinational farm business and added some tractors to his line.

Bottom: 1930 Vickers-Aussie tractor in the Whiteman Museum collection, WA. (Photo courtesy *The Old Machinery Magazine*, December 1997)

Plate 6 (between pages 88 & 89)
Top: Tail-end Charlie, a Chamberlain Champion specially modified for high speed road work, is in working condition in the Whiteman Museum in WA.

Bottom: Tail-end Charlie could reach 68 mph. This tractor followed several round-Australia rallies, having to cover a distance of 11,140 miles to complete the circuit in 19 days in the 1957 Mobilgas Trial.

Plate 7 (between pages 88 & 89)
Top: Author's photo of one of the Hart-Parr Australian Special Model 12-24Es. This well-restored tractor was at Queensland's Jondaryan Woolshed Museum.

Bottom: When you visit the Jondaryan Woolshed be sure to inspect one of the most unusual tractors ever to be seen. Irvine's 37-ton Road-Burner was a mobile kiln that was used in the late 1920s to bake clay roads to make them trafficable in wet weather in outback Queensland. A team of 23 accompanied the machine, which worked 24 hours a day on the job and burned whatever fuel was available. It travelled at 220 hours per mile! (Author photo)

Plate 8 (facing page 89)
Top: KL Bulldog 40 HP crude oil tractor. The Bulldog was restored by Bert Den Hartog of Camden, NSW, who provided the photo.

Bottom: KL Bulldog driven by Steve Muscat of Riverstone, NSW, helping out a stalled Field Marshall during TOMM's Inaugural 1998 Wingham Classic Tractor Trek. (Courtesy *The Old Machinery Magazine*, April/May 1998, back cover)

Plate 9 (facing page 112)
Top: Tandem Chamberlain rig built by South Australian Mallee farmer, Jeff Morgan. Powered by two six-cylinder GM 671 naturally-aspirated diesels, the machine boasts six sets of duals in an articulated chassis with rear axle bogied. This is a working tractor that has been used for crop establishment for 20 years at Karoonda near Lamaroo. Jeff Morgan states that building this triple-axled tractor from Chamberlain Super 70 basics has cost him less than $12,000 for a one-of-a-kind 450 HP machine. (Courtesy *The Old Machinery Magazine*, back cover April/May 2005.) This is the machine on the front cover of the book.

Bottom: The Chamberlain light tractor was designed by and two prototypes built under Bob Chamberlain's direction in 1959 with the idea of competing with the Ferguson. The 'Light' was powered by a standard Holden engine. Economic circumstances prevented commercial production. One of these tractors is preserved by the Museum of Victoria. (A.H. Chamberlain provided the photo).

Plate 10 (facing page 113)
Top: Model 4480 (98 HP) of the Chamberlain Sedan series. This tractor outsold all others in its power class in Australia for five straight years. This model continued the Australian company's two-wheel-drive rugged tractor design style, with 12 forward speeds and four reverse gears; it was fitted with a John Deere US-built engine. It was one of the last to carry the Chamberlain name.

Bottom: Howard's DH22 tractor was launched around 1930 and was manufactured for 30 years. This DH22 was restored by Ross Sempf, Gatton Queensland. (Photo Courtesy The Old Machinery Magazine, February 1997)

Plate 11 (facing page 128)
Top: Waltanna Model 44-400 made by Victorian tractor builder, James Nagorcka. It was fitted with a Gason cab in 1985.

Bottom: The Australian-built Ford FW25 four-wheel-drive tractor which first appeared in 1986 (163 engine HP). It was built by Waltanna at Tarrington, Victoria, for Ford Australia. (Author's collection)

Plate 12 (between pages 128 and 129)
Top: Geoffrey A. Williames at Warragul, Victoria, launched a series of low-CG 4WD machines initially as the Supercat from 1973 on, and was followed by others who came out with similar concept designs. The Austrak appeared between 1978 and 1986.

Bottom: Williames Hi-Tech International in Warragul, Victoria, made this track-mounted self-propelled tea harvesting machine with locally-built silent track — a wire rope-reinforced urethane belt with bolted-on grousers for traction. It could harvest 15 tons of tea leaf from 30 acres in a day. Nowadays, the T750, an eight-wheel-drive model, is offered to order.

Plate 13 (between pages 128 and 129)
Top: Baldwin Model DM600 broadacre rig. This four-wheel-drive 600 HP machine was one of the Baldwin tractor series produced in Sydney between 1979 and 1988.

Bottom: Horwood Bagshaw's Acremaster 4WD between 1986/87.

Plate 14 (facing page 129)
Top: Richter Self-propelling Palm-Fruit Tree Harvester.

Bottom: Richter Hydrostatic Tool Carrier Model TC22 at work in vegetable transplanting mode. (Source: Richter photos)

Plate 15 (facing page 144)
Top: Alfarm-Mizzi self-propelled sugarcane harvester for green cane harvesting. Only a handful were ever sold.

Bottom: This cane harvester stamp was issued in 1991. Very few Australian stamps have been issued showing tractors or farm machinery, maybe three in total, unlike some countries that have had essentially no farm equipment industry, like Mongolia and New Zealand.

Plate 16 (facing page 145)
Top: Goldacres Trading Pty Ltd in St Arnaud, Victoria, are manufacturers of trailed spraying equipment and self-propelled boom sprayers. Goldacres make four SP models. The 4WD SP model shown has a boom 33 feet wide that is air-suspended.

Bottom: The Dingo mini-loader is made by Dingo Mini-Diggers Pty Ltd in Dalby, Queensland, and exported widely. Model K9-3 is shown on the front cover. Here is a purpose-built remote-controlled model K9-4 for situations where conditions are too hazardous for the operator's direct presence. Launched first in Australia, the market for mini-loaders has expanded enormously and now there are at least 12 makers involved in the United States, as well as the two Australian manufacturer/licensees, Dingo and Kanga, both based in Queensland.

INDEX